Fast and Slow

Fast and Slow: Methods for Timely HCI and Interaction Design Research explores the dynamic interplay between rapid and reflective, or long-term vs. short-term research in human–computer interaction (HCI) and interaction design research. In an era where fast approaches to technical research are foregrounded, this book provides a critical examination of the temporalities at play in various research approaches – from long-term empirical studies to rapid prototyping. Ultimately, it asks the fundamental question of how to make timely research contributions and how to plan research projects to be timely in terms of impact.

Drawing inspiration from fast and slow thinking, design thinking, and our ever-changing world, this book contrasts "fast" approaches – such as "quick-and-dirty" ethnography, prototyping, and AI-driven automation – with "slow" methodologies that emphasize ethnographic studies, longitudinal research, and participatory design. By working across the fast and slow approaches to research, the book helps researchers and practitioners navigate the trade-offs between efficiency and depth, rapid results, and reflection. Ultimately, the focus of the book is on timely research contributions.

With a focus on research approaches, this book presents cases, methodological insights, and theoretical discussions that foreground the temporalities at play in HCI and interaction design research. It questions whether the rush to develop, iterate, and deploy can sometimes obscure critical insights about human behavior, emerging phenomena, ethical considerations, and long-term impact.

Whether you are an HCI researcher, UX practitioner, or technology strategist, *Fast and Slow: Methods for Timely HCI and Interaction Design Research* offers a fresh perspective on how to plan and carry out HCI and interaction design research – over time.

Fast and Slow

Methods for Timely HCI and Interaction Design Research

Mikael Wiberg

CRC Press
Taylor & Francis Group
Boca Raton London New York

CRC Press is an imprint of the
Taylor & Francis Group, an **informa** business

A CHAPMAN & HALL BOOK

Graphical illustrations – Viktor Wiberg

First edition published 2026
by CRC Press
2385 NW Executive Center Drive, Suite 320, Boca Raton FL 33431

and by CRC Press
4 Park Square, Milton Park, Abingdon, Oxon, OX14 4RN

CRC Press is an imprint of Taylor & Francis Group, LLC

© 2026 Mikael Wiberg

ISBN: 9781032381541 (hbk)
ISBN: 9781032381596 (pbk)
ISBN: 9781003343745 (ebk)

DOI: 10.1201/9781003343745

Typeset in Minion Pro
by Deanta Global Publishing Services, Chennai, India

Contents

Preface

R ESEARCH TAKES TIME. AND research contributions need to be timely – it is part of how a contribution is defined – it presents something new in relation to a research problem and in relation to an established body of related research. As such, research is relational and temporally situated. Research is also not just temporally situated, but also temporally motivated and restrained. Sometimes there is an urgent research problem that needs to be addressed, and sometimes there is just too little time when some longitudinal studies would be the preferred approach. Over time, research problems emerge, new methods are developed, and progress is made – through new studies, explorations, insights, discoveries, and new ways of seeing things. Through collective efforts – ranging from research collaborations, projects and research programs, to seminars, peer reviews, and presentations – research is always in motion and particular research projects are accordingly always situated and defined in relation to these collective movements – again, over time. And over time, new research problems are identified, old problems are revisited, and sometimes even old theories and conclusions are challenged. Sometimes there is a research breakthrough, where huge steps forward can be taken, and sometimes researchers struggle for decades with the same research problem. In short, this dynamics of how research is conducted is something that plays out over time. Given this, the pace of research is not constant, nor is it possible to talk about "one pace" or "one research process." Still, very little has been said about the different temporalities at play during the research process, the importance of being timely when selecting and addressing a research problem, and the importance of being timely to make a solid research contribution. This book is about these temporalities at play, about the "landscape of temporalities" that surround, influence, motivate, and restrain research – particularly in interaction design research.

If taking a step back, we can notice how the two concepts of "time" and "temporality" are often taken for granted. We use these words in our everyday language when we say "well, it takes time," "let's wait and see," or "let's wrap this up before it's too late," which speaks to how things evolve over time, or that it is about time to finish something. Somehow, there seems to be a window in time where a timely research contribution can be made. Still, although we use these terms in our everyday language, they are also quite complex concepts to explore. As I illustrate in this book, these notions of time, temporality, pace, rhythm, etc., can work as a vocabulary, perspective, and lens that allow us to see how things unfold in research processes – over time.

On the one hand, time can be thought of as a concept that we either experience or measure. For instance, how we might feel bored over longer periods with no activities, how we might feel when we are behind schedule, when we are about to miss a deadline, or when we lose track of time because we are immersed "in the moment." We can also use time as an indicator of how well we perform. We can formulate time plans, KPIs, GANTT charts, deadlines, etc., and through these time-based tools, we can estimate and then measure or evaluate how long some activities have taken, how far we are from a deadline, if we need to speed up, or how much time we have left. In short, time affects our behavior, activities, priorities, and how we make plans. But on a higher level of abstraction, we can also think about time and temporality as a fundamental construct that shapes our lives – from clocks, calendars, and to-do lists to deadlines, timetables, and ways of doing several things – one thing at a time, in a certain order over time, or at the same time, i.e., multitasking. Time, and how we relate to and organize ourselves around it, shapes our everyday lives, and as I illustrate in this book, it also shapes the research we do, and how we organize our research processes – explicitly or implicitly, no matter if we want it or not.

In the research areas of human–computer interaction (HCI) and interaction design (IxD), time and temporality play a crucial role in structuring research – over time. It influences the problems we address, the methods we choose, the technologies we focus on, the theories we pick, the timelines and deadlines we set, and how we adjust our research ambitions in relation to how much time we estimate that we have – explicitly in relation to project goals, work packages (WPs), and project duration, and more implicitly in relation to our understanding of how the research landscape is moving and changing, and if we will manage to make a timely contribution to that movement. As I illustrate in this book, different paces are at play, with some processes moving quickly, while other activities and processes are moving more slowly – i.e., fast vs. slow research processes and approaches. As researchers, we are constantly making estimations of these paces, ranging from practical matters such as how urgent a research problem is, or how long a research project should be, to more overarching matters – such as if what we are studying is a new topic, a novel approach, a trend, or an escalating problem. Across all these cases, time is a central concern and a governing principle for any research project. No matter how you spin it, time will always be a factor.

In this book, I explore, contrast, and propose ways of unifying fast and slow approaches to interaction design research. Related to these perspectives on "fast" vs. "slow" approaches to design research, I have of course been inspired and influenced by the pioneering book *Thinking, Fast and Slow* by the Nobel Prize winner, Professor Daniel Kahneman (Kahneman, 2011) – a book that has already been cited more than 53,000 times. I have also been inspired by the growing body of research on what has been labeled "the slow science movement": see for instance the book *The Slow Professor* by Berg and Seeber (2016), and the book *Another Science Is Possible: A Manifesto for Slow Science* by Isabelle Stengers (2016). These books examine an alternative, slower temporality for conducting research, and it is interesting to relate these alternative approaches to the temporalities at play in interaction design research. I have also been inspired by the growing body of related research that looks into fast or slow approaches to research or work that considers the temporalities at play when planning or conducting short or longitudinal research, for instance, the work by Sauer (2018) on fast and slow moral thinking, and the work by Kannengiesser and Gero (2019). On fast and slow design thinking see Earley (2017) and Keulemans (2021) and on fast and slow design see Frohlich (2015). I have also been inspired by work on fast and slow innovation: Gasparin et al. (2020) on slow design-driven innovation and Goldsworthy et al. (2018) on "circular speeds" and fast and slow design approaches. I have also been inspired by McCracken's (1988) book *The Long Interview*, Croskerry et al. (2014) on fast and slow decisions, Cal Newport's book *Deep Work* (Newport, 2019), and Dorie Clark's book *The Long Game: How to Be a Long-Term Thinker in a Short-Term World* (Clark, 2021). Finally, I have also been inspired by more general literature on time, temporality, and the experience of time, for instance, the book *Fast and Slow: Design and the Experience of Time* by Hayward (2016).

Understanding and foregrounding the different temporal paces at play in HCI and interaction design research is important because it allows us to navigate and make plans in a rapidly changing research landscape. At the current moment, there are several drivers that push for fast approaches in our field, including new and emerging technologies (at the current moment in time, this includes, for instance AI – artificial intelligence, and IoT – the Internet of Things), research funding, call for papers, and bibliometrics. On the other hand, there are calls for slow approaches to address complex and hard-to-solve research problems. In this context, fast approaches have often been seen as "relevant" and well-aligned with tech paradigm moves and the buzzword-driven landscape surrounding digital technology research and development. On the other hand, slow approaches have typically been associated with "rigor" and the formulation of new paths, paradigms, theories, research directions, and new orientations and movements in a field of research.

Traditionally, HCI and IxD research have focused on fast solutions (most recently discussed as the "solutionism" movement in HCI. For a discussion on this, see for instance: Cunningham et al. (2023), Adamu (2023), Aanestad (2023), Blythe et al. (2016), and Oulasvirta and Hornbæk (2016). Along this line, we find a set of approaches for going "fast" in the research process, including methods like brainstorming (see e.g., Wilson, 2013; Sutton & Hargadon, 1996); Besant, 2016; and Putman & Paulus, 2009), design briefs (Ryd, 2004), quick and dirty ethnography (Twidale et al., 2014), and rapid prototyping. However, there has been a recent turn in the field of HCI toward "big debates" including topics such as race, gender, climate, sustainability, and social justice, to just mention a few of these more complex topics. In short, a set of issues and problems that cannot really be quickly solved, and things that cannot just be addressed once and then call it solved. On the contrary, these issues call for long-term and slow changes, and these topics even call for fundamental and systemic shifts. Such shifts take time.

Ultimately, as I illustrate in this book, the choice between, or the combination of, fast and slow methodological approaches in HCI and interaction design research is a matter of paying attention to the urgency, complexity, and scale of the research problem – in relation to how a particular field of research develops – over time. Large-scale research problems require collective efforts and coordination, often involving large interdisciplinary research teams and long-term funding. This often necessitates slower, more rigorous approaches to understanding and addressing large-scale challenges. On the other hand, research contributions are typically made through the documentation and publication of timely studies in relation to how a research field develops over time. Accordingly, it becomes a practical and everyday challenge for any researcher who wants to make a difference – to simultaneously address an urgent and large enough research problem, while managing to break it down into smaller (and shorter) studies and aim for timely contributions.

So, how do you as an individual researcher navigate this ever-changing temporal landscape? How do you plan your research projects so that you can make timely contributions? And ultimately, how do you contribute to where the field of research should go next, or where it needs to be, in say 10 years from now?

This is what this book is about.

–"Go fast, Go slow" – Enjoy!

About the Author

MIKAEL WIBERG IS FULL Professor in Informatics at Umeå University, Sweden. He has held positions as Full Professor in Interaction Design at Chalmers University of Technology and as Chaired Professor in HCI at Uppsala University. Wiberg's research is within the area of HCI/interaction design research, and he has published his research in top journals including *Design Issues, Design Studies, International Journal of Design, ToCHI – ACM Transactions on Computer–Human Interaction, the Human–Computer Interaction* journal, *Interacting with Computers.* He is also co-EIC (Editor-in-Chief) for *ACM Interactions.*

CHAPTER 1

Fast and Slow Design Research

INTRODUCTION

How we step-by-step move forward in interaction design research, and how we motivate and describe these steps, is a matter of method. And, at what pace we move forward through these steps is a matter of the temporalities at play – no matter if we move rapidly through the research process, or if we are slowly and steadily moving forward.

In this book, we examine the temporalities at play in interaction design research. We do this through an examination of how design ambitions and research efforts come together in design research. This examination is grounded in an analytical approach to how design research has developed over the last decade. The book is also rooted in my own experiences from doing design-oriented research over the past 25 years – in particular in the research areas of HCI (human–computer interaction) and interaction design research. While there is a growing body of research on design, design methods, and research approaches to design [for instance, RtD – Research through Design (see e.g., Zimmerman & Forlizzi, 2014; Bardzell et al., 2016; Godin & Zahedi 2014; and Giaccardi, 2019)] less is said about *the temporalities underpinning and influencing interaction design research processes.* In this book, I suggest that an understanding of these temporalities at play is of essential value and importance for a number of reasons: 1) for the strategic planning of interaction design research projects, 2) to understand the relation between design and research processes, and how these come together in design research, and 3) to understand how to align basic and applied design research agendas in practice. Taken together, these three aspects are crucial for doing timely research, and accordingly, it is essential for making timely research contributions in the area of HCI/ interaction design research.

On an overarching level, this book is about the processes, rhythms, and paces of interaction design research. It is about the *practice of doing* timely research, and it is about *research strategies* for how to aim for long-lasting impact – through interaction design research. It is about how to combine a focus on the most pressing issues and challenges, while aiming for long-term progress and long-lasting impact. Historically, HCI and interaction design research have been very future-oriented, design-oriented (Wiberg, 2014; Wiberg & Stolterman, 2014), and problem-solution-oriented (Oulasvirta & Hornbæk, 2016). Solutions and novelty have been criteria to measure if a research project is producing valuable results – or not. However, this situation is now changing, and there are even papers that criticize HCI for pushing a "solutionism" agenda in favor of more long-lasting, or more long-term impact. In short, it is about going fast – and finding solutions to problems, or going slow – and working on more complex (and even wicked) research problems.

There are of course several books already on design methods and design thinking, and there are books available on particular areas of design practice, for instance, books on interaction design (e.g., Lowgren & Stolterman, 2007). Still, there is a lack of literature that specifically addresses the temporalities at play in interaction design research. This book serves as an important contribution to this need. If we look at the existing strand of literature there are a set of books that are of particular relevance here, although none of these books focuses on the temporalities of design research. For instance, the book *The Design Way* by Nelson and Stolterman (2014) provides an in-depth perspective on design, design theory, and design as an approach. This is an important book as it provides a foundation for understanding the nature of design and design processes. In close relation to this book, I would also like to acknowledge the book *Making Design Theory* (MIT Press) by Redstrom (2017). This book is important as it describes how theorizing can be made in design research, how applied and basic research is interrelated in design research, and how design researchers can work along "programs" in their research efforts.

In addition, we have the book *Transmissions* (MIT Press) by Jungnickel (2020). This book looks at the outcomes, presentations, and dissemination of design research. According to the author of this book, "Transmission is the research moment when invention meets dissemination—the tactical combination of making (how theory, methods, and data shape research) and communicating (how research is shown and shared)." This is a book of great importance and value for understanding the processes surrounding design research, in particular toward the end of a research process where this transmission is not just about showing and sharing research results, but about presenting these results in a timely manner. In short, I think this book adds to these other important books by providing a particular focus on the temporalities surrounding and underpinning interaction design research processes, and how to align those internal research processes with what is happening in practice and what is happening in our surrounding society – at a given moment in time.

We can also notice here how this book on *timely interaction design research* is well-aligned with a number of books that point to the dynamics of our everyday, and accordingly how we need very dynamic design research processes that are well-suited for and capable of following and contributing to this ever-changing landscape of our everyday. Here, I would like to mention the classic book *Beyond the Stable State* by Donald Schön (1971), the book *Liquid Modernity* by Zygmunt Bauman (2013), and the book *The Necessity of Friction* by Nordal Åkerman (2018). These three books are all about this ever-changing dynamics in our everyday life. On the level of the individual, and how we need to cope with an ever-changing everyday full of challenges, changes, and interruptions, I am also thinking about the book *The Distracted Mind* by Adam Gazzaley and Larry D. Rosen (2016). Still, and while all of these related books examine and foreground the dynamics of our ever-changing everyday, none of these books address this from the perspective of how to plan and carry out interaction design research that is timely in relation to such unstable, uncertain, and dynamic conditions, circumstances, and surroundings. This is where this book on timely approaches to design research contributes with a perspective on how to plan and work with fast and slow approaches to interaction design research in relation to this dynamic object of study. In addition to this, if we look at how this book might contribute more broadly to the field of interaction design research, I would say that this book brings together these perspectives offered by the existing literature (i.e., the books on design/design theory, the books on the dynamics and ever-changing nature of our contemporary society, and books that have highlighted the importance of alternative temporalities for doing research).

By bringing these perspectives together, I see this book as something that hopefully offers a set of important contributions in relation to the growing body of interaction design research that specifically addresses ways of approaching design research (e.g., design research methods) and ways of theorizing design research. In addition to this, I think that this book might serve as a thought piece for design researchers interested in reflecting upon the temporalities, rhythms, and paces that form, influence, push, or even restrict research projects. It might also serve as a guiding framework that can make design researchers better – both in terms of planning new projects (a process perspective) and in terms of making timely contributions (an outcome perspective). Finally, I hope that this book can provide some guidance on thinking about interaction design research as a matter of work that is oriented toward the formulation, articulation, and realization of *prospects for alternative futures.*

While my own research over the past 25 years has mostly been in the area of interaction design/HCI (human–computer interaction), I think that this book might be relevant for almost anyone interested in design research approaches and ways of doing timely research through design. As such, I see the audience of this book being designers, design researchers, HCI and design students, PhD students in HCI/interaction design, interaction designers, researchers in HCI – or in any other area where design, proposals, recommendations, or implications are part of the method, approach, or foundation for doing research and for moving forward through timely contributions that make our world a better place – for all of us!

Some Initial and Fundamental Questions – on Fast and Slow Interaction Design Research

When do we design? *What* does it mean to do timely design research? And accordingly, *how* should we plan our interaction design research projects? Indeed, these are concerns for the temporalities of interaction design research. When we plan and carry out interaction design research projects, we need to account for and deal with the temporalities of research processes. We need to do our projects on time, in time, and within deadlines, and we need to be timely with our results and contributions. Still, when reviewing the existing literature on interaction design research, very little is said about these temporalities. So, what are *the rhythms of design* and the *paces of research*? And how can these two come together in interaction design research projects? On an overarching level, this book is about the processes and paces of interaction design research. It is about doing timely research, and how to aim for long-lasting impact. It is about how to combine a focus on the most pressing issues and challenges while still aiming for long-term progress and impact.

In this book, interaction design research is examined from a temporal perspective. It includes both the planning of HCI/interaction design research projects, as well as with internal and external factors influencing the temporality of these processes. For instance, we might need to *move fast* to find solutions to pressing problems, or we might need to be first to contribute to a fast-growing strand of research. In other processes, we might need to dig deep and *move slowly* to fully understand and analyze a particular setting. We might also need to be future-oriented in the planning of our research projects, ensuring that our contributions toward the end of the project will be relevant at a future moment in time. Certainly, interaction design research moves forward at multiple paces, making it is accordingly a relational concern to ensure that the things we do now will be relevant not just in the present, but also at later stages and for longer periods of time.

In addressing *the rhythms and paces of interaction design research*, I will, in this book, propose and use the notions of "fast" and "slow" as a vocabulary that allows for an examination of these rhythms and paces of design research. Here, I define fast and slow as relational notions. A research process is *fast* if it is more rapid than how the research problem changes or develops. If you are fast, you can address it while there still is time, or while the research problem still exists or is relevant. Likewise, being *slow* is also relational. It can be that you are too slow, so you arrive at a solution but it is too late. It can also mean that you decide not to address the urgent research problems, but instead try to tackle or at least understand the more fundamental or underlying problem. This will take more time and accordingly lead to a longer and as such slower research process. Accordingly, I will rely on an understanding of "fast" as being about things that move forward rapidly, in quick succession, or ahead of an estimated time, deadline, plan, or schedule. Further, I will view "slow" as a different pace, although not necessarily a negative one (e.g., "being delayed"). Instead, I approach "slow" as a process that has qualities that might be necessary for moving forward at a rapid pace, through rapid progress and actions. Here, slow processes that have involved reflections and activities that have laid a solid foundation might provide the grounds necessary for dealing with urgent research problems – at a later stage in time.

In our language, "fast" is typically associated with "action," "achievements," and "progress," whereas "slow" is typically associated with "reflection," "analytical," and "systematic" approaches. In interaction design research processes, we typically combine these two. We seek, we do, we iterate, we evaluate – we *act*, and we *reflect*. This is one way through which the fast and slow approaches come together in interaction design research practice. But we also have similar paces in research. We might be stuck with a research problem for a very long time, and suddenly we have a breakthrough. It is hard to predict *when* research demands further analysis, and *when* there will be a sudden shift that enables a rapid movement forward. Again, this is an example of the temporalities at play. So what do I then mean when we talk about paces and rhythms in this context? Well, *pace* is the overarching speed of the research project. Again, it is relational in many ways. The pace might be high in comparison with the research plan, or the pace might be slow because of challenges with getting access to data, etc. The notion of rhythm is related to this pace. On an overarching level, the pace might be slow, but there might be recurring moments with a higher pace. For instance, if the plan is to run a set of workshops as part of a project, then these workshops might run smoothly and according to the project plan, although the project as a whole is moving forward at a quite slow pace. As a researcher, it is accordingly important to not only understand the overarching pace of a project, but also to have a good understanding of fast vs slow activities, temporal bottlenecks, and time-consuming moments. With such knowledge, it is easier to plan research projects over time and be in a good position to make timely research contributions.

On a related note, it should be acknowledged here that I do not see "fast" as being sloppy or not being rigorous enough, and I do not see "slow" as a matter of being lazy, delayed, or too late. Instead, the two temporalities of *fast* and *slow have* their unique qualities, and in order to do timely design research, it is essential to understand how these different paces – these different temporalities of design research – are interrelated and interdependent.

On Timely, Recurring, and Eternal Research Questions

Beyond these two notions of moving fast and moving slow, there are also different types of research questions. Some questions are formulated to address urgent problems, while other questions are formulated to address slower or less urgent problems. But there are also questions that will probably stay with us for a very long time, for instance, questions concerning sustainability. As you will see in this book, I will move from thinking about fast vs. slow research to discussing this in relation to basic and applied research. Here, the idea is that applied research is probably more suitable for a fast approach, and vice versa with basic research and a slower approach to research. But beyond these two alternatives, there are probably research questions that will stay with us – some kind of eternal research questions. Here, I will introduce this idea and then return to this topic toward the end of this book.

For sure, every field of research has its own set of "eternal," or at least recurring research questions. In our field, the field of human–computer interaction/interaction design research, we seem to revisit the question of how to keep up with the pace of technological developments, and sometimes we even seek to add to the forefront of technological advancements, while paying attention to the societal adoption of new digital technologies. At the same time, we try to make sure that we contribute to something more fundamental or long-lasting, i.e., to generate new knowledge and new theories about these processes.

As a field of research, we seem to struggle with how to conduct research that is *fast enough* to keep up with the current pace, and at the same time, engage in interaction design *thoughtfully* (Lowgren & Stolterman, 2007) *and thoroughly enough* to meet the scientific criteria of rigor. As a field of research, we need to contribute with concrete results and solutions (new gadgets, tools, and digital services), something that has sometimes been referred to in the literature as "ultimate particulars" (Nelson & Stolterman, 2014), while at the same time making sure that we generate abstractions, generalizations, and that we theorize (Stolterman & Wiberg 2010; Wiberg & Stolterman, 2019; Wiberg, 2018). It is as if we both want fast and concrete results that we can show, demo, and exemplify with, while at the same time coming back to the fact that we also want to make conceptual advancements in our field of research.

In adjusting our research agenda in relation to the pace of technological developments and changes in social practice, our field has been occupied with the latest technologies and with what's "new and novel" (Wiberg & Stolterman, 2014). We have also developed a keen eye toward innovation, novelty, and the next big thing. When we report our research findings, we nowadays typically strive to include "implications for design" (Dourish,

(2006), and we are excited and explicit about our design-oriented and future-oriented research agenda. We want to make an impact, and we want our research to matter – right now, and right away! To be relevant, timely, theoretically grounded, and future-oriented seem to be four central cornerstones for our field.

From a temporal perspective, we do our research "here and now," but at the same time we want it to be about the coming technological shift, the next big thing, e.g., next wave of computing, or future use scenarios. We want to understand and describe the present, but we also want to be normative and suggest where we should go next. In a sense, we are a forward-looking field rather than a field that only seeks to understand the current moment, and we seem to be occupied with new technologies and what futures these new technologies can enable for us (Bell & Dourish 2007). We are not only future-oriented, but in this strive, we are interested in the new and novel (Wiberg & Stolterman, 2014). For instance, HCI's most established conference, the ACM CHI conference, is explicit on how researchers should present their work in terms of its novelty. There is an embedded assumption here that we should not just focus on the current situation or the most recent technologies, but we should also draw conclusions about the next steps forward. We should be future-oriented, and we should seek to target, or even establish, the next phase or the next "wave" of computing. Indeed, our field has, over the last 20 years, been expanding across such waves, and in multiple new directions (for further discussion on these waves see e.g., Bødker, 2006, 2015; Kaptelinin et al., 2003, April; Goh et al., 2017; Frauenberger, 2019; Ashby et al., 2019, November; Homewood et al., 2020, July; Lopes, 2021; and Wiberg & Stolterman 2021). In addition, the *ACM Interactions* magazine has just recently launched a new publication format called "waves" devoted to articles that report on new and emerging topics and strands of research in our field.

It needs to be stated here that we are not only future-oriented, but we are also a design-oriented field of research. We are in that sense breakthrough-oriented, and we are expected to deliver timely research contribution – the CHI conference even has a submission category for "late-breaking results" as to ensure that the most recent results will reach out as soon as possible. At the same time, we are design-oriented, both in our approach and in our strive, as well as in terms of theoretical advancements. We want to expand on design research as a body of research, and we want to make theoretical contributions to the field of HCI. With this as part of our research agenda, we have pressure on us to systematically develop our theories, be critical, contribute to the existing body of research, and generate insights and new knowledge with long-lasting implications. In short, we are expected to be both *fast* and *slow* in how we conduct our research. This is important, as these expectations shape how we conduct research in our field.

In this book, I focus on these different goals and temporalities of HCI and interaction design research. I discuss how to navigate the different paces surrounding our research projects, and how to do interaction design research that is both timely and long-lasting. For sure, our field has managed to do this and we actually have experience from doing it with every interaction design research project we conduct. Still, very little has been written about it from a research process perspective. As such, this book contributes to the existing literature on methods in HCI (human–computer interaction) and interaction design research, and it provides a reflective account that will hopefully help new scholars entering our field, particularly in understanding how to combine both fast and slow activities in research plans to be both timely and relevant, as well as critical, thoughtful, and grounded. This is at least the intention of this book.

TAKING A STEP BACK TO REFLECT – INTERACTION DESIGN RESEARCH, AND ITS FAST AND SLOW METHODS

On an overarching level, one can say that interaction design research is a field concerned with people, activities, context, and technologies (sometimes referred to as the "PACT" approach Benyon, (2014)). One might also say that it is about the design and evaluation of interactive products and systems, such as software applications, websites, and smart devices (see e.g., Wania et al., 2006; and MacDonald & Atwood 2013). For sure, usability testing has played a central role in the field of human–computer interaction, where usability scales such as the one proposed by Brooke (1996) has been frequently applied and referred to. At the current moment, this proposed usability scale has more than 20,000 citations. As technology continues to permeate our everyday lives, the importance of good interaction design has not only grown, as it plays a critical role in determining the usability, effectiveness, and satisfaction of these products and systems, but it has become fundamental since these technologies are now so entangled with our everyday lives that any attempt to separate the two is meaningless, if not impossible. Some researchers, e.g., Frauenberger (2019) has referred to this as the fourth wave of human–computer interaction – or entanglements – a stage where we are not only using and surrounding ourselves with these digital systems, but where these systems also to a large extent define who we are, what we can (and cannot) do, how we understand and relate to the world around us, and how others perceive and interact with us. In short, we are existentially entangled with these digital systems in the fourth wave of HCI.

If now taking a step back, we should acknowledge how design research, as the broader research domain surrounding interaction design, is a field devoted to a systematic and rigorous approach to understanding the needs, motivations, and behaviors of people in order to inform the design of products, services, and experiences (Hartson, 1998). It is typically described as an interdisciplinary field (Blackwell (2015); Mackay & Fayard (1997); and Rozanski & Haake (2003) that combines elements of human-centered design, social sciences, and engineering to create a holistic understanding of the design context. The goal of design research has been described as generating insights that inform and inspire the design process, leading to innovative and effective solutions (Packer and Keates, 2024). Further on, it has typically been described as being done through an iterative design process that involves gathering and analyzing data from a variety of sources, including interviews, observations, surveys, and experiments (Dow et al., 2005; Myers, 1994). There are also many different methods of design research, including qualitative and quantitative approaches, that can be used to gather data from various stakeholders. These include user interviews, focus groups, usability testing, and field studies, among others. One key aspect of design research is the ability to identify and prioritize the needs of the target audience. This requires an empathic understanding of the users, their values, their needs and activities, and their motivations. It also requires an understanding of the broader context in which the design will be used, including cultural, social, and environmental factors. From the early ideation stage through the final evaluation of a product or service, design research can be used at any point in the design process. It is a crucial tool for designers who want to build solutions with effect and meaning that cater to the demands of their users.

Design research can also be part of a traditional design process (Jones, 1992), where it assists professional designers in the early stages of the process by defining the issue they are seeking to address and locating the important parties involved (Svetel et al., 2018; Collins et al., 2016; Zimmerman et al., 2007). Insights about the users and their needs can also be uncovered, which can guide the creation of personas (Chang et al., 2008) and user journeys (Benford et al., 2009; Bascur et al., 2019; Tan, 2023). Design research can also be used to evaluate and validate concepts and prototypes as the design process develops. This can be accomplished using techniques like focus groups (Adams & Cox, 2008; Rosenbaum et al., 2002), user interviews (Blandford et al., 2016; Raita, 2012; Kuniavsky, 2009), and usability testing (see for instance Nielsen, 1992, 1994; Lewis, 2012; Shneiderman, 2000, and Issa & Isaias, 2022). Designers can also evaluate and enhance their designs with the help of the comments and information acquired through these methods and techniques.

In the final stages of the design process, design research can be used to evaluate the effectiveness and impact of the final product or service. This can involve gathering data on usage patterns, user satisfaction, and the overall user experience. Overall, the role of design research in the design process is to provide a solid foundation of knowledge about the users and their needs, which can inform and inspire the design process, leading to innovative and effective solutions.

Related to these contextual factors, the existing strand of research methods, and this recurring focus on users and users' needs, we should also acknowledge a recent movement in the field of design research that seeks to decentralize the human in these processes. Over the last few years, this has been labeled more-than-human centered design (Coulton & Lindley, 2019; Wakkary, 2021; Camocini & Vergani, 2021; Poikolainen et al., 2022; Giaccardi et al., 2024), and by de-centralizing the human this approach suggests that we need to focus on larger (eco-)systems to tackle global sustainability challenges. In a sense, this recent shift from *human-centered* to *more-than-human centered design* research serves as a good example of this combination of fast and slow movements in our field of research. On the one hand, there have been lots of rapid research projects conducted where people have looked for human-centered solutions to various problems. At the same time, there has been a slower movement underneath the fast-paced research, a movement where researchers have started to question if these solutions should only serve human needs, and the environmental footprint associated with pushing a human-centered research agenda. With the turn to more-than-human approaches, it is on one level a slow turn in the field, where researchers are seeking an alternative foundation for doing and evaluating research outcomes, but it also comes with an additional set of more rapid research where researchers are then conducting more-than-human centered interaction design research projects in attempts to explore what this turn could mean in practice.

Design research can also be used to assess the success and impact of the finished product or service in the last stages of design. Data collection on usage trends, user happiness, and the overall user experience may be necessary for this. Overall, design research's function in the design process is to lay a strong foundation of understanding about the users and their demands, which can guide and inspire the creation of innovative and efficient solutions.

If we now go into detail about the different methods available in the area of interaction design research, we can see that there are several approaches to interaction design research, which can be broadly considered as fast or slow approaches. Fast approaches are characterized by their focus on rapid iteration and prototyping, often involving user testing at various stages of the design process. Slow approaches, on the other hand, tend to be more reflective and iterative, with a greater emphasis on understanding the context and needs of users over the long term. In this book, I explore both fast and slow approaches to interaction design research, examining their strengths and limitations and considering how they can be effectively combined in practice. In doing so, we also delve into the social science literature on topics such as user-centered design, usability engineering, and participatory design, as well as the specific field of human–computer interaction (HCI).

At the current moment, there is a growing "slow research" movement in HCI. This movement is not about "slow science," but rather about the development of a new sensibility that allows for alternative sets of research questions. As we slow down, we can start to see things differently, and it allows us to redirect our focus and our efforts. And as we change perspective from a design-oriented or solution-oriented approach that foregrounds how to quickly solve problems [sometimes referred to as "solutionism" (Cunningham et al., 2023)], we start to engage with questions that might not have any clear answers, simple solutions, or "quick fixes," for instance, ways of addressing social (un)justice, gender inequalities, and the global sustainability challenges. In short, it is a move toward "post-growth" approaches to HCI and interaction design research (Sharma et al., 2023).

Coming back to the definitions at the beginning of this chapter, I would in this context say that the "fast" vs. "slow" approaches to design research refer to the speed at which data is collected and analyzed, as well as the depth and breadth of the research – again in relation to the real-world research problem, and in relation to the research plan for the project. To provide an overview of these two approaches, one can say the following:

Fast design research methods are guidelines for those studies that can be conducted quickly and with minimal resources, and often involve collecting data from a small number of sources. Examples of fast design research methods include online surveys, quick user interviews, rapid prototyping, and lightweight usability testing. These methods are useful for gathering initial data and testing basic assumptions, but may not provide as much depth or nuance as slower methods.

Slow design research methods, on the other hand, are guidelines for more time-consuming and resource-intensive studies, which often involve collecting data from a larger number of sources. Examples of slow design research methods include in-depth interviews, focus groups, and longitudinal field studies (Åhlström & Karlsson, 2010; Kjærup et al., 2021; Lucero et al., 2021). These methods can provide a more comprehensive understanding of the users and their needs but may take longer to conduct and analyze. For sure, if addressing systemic problems, there is a need for such in-depth studies.

In this book, I pinpoint how both fast and slow design research methods have their place in the design research process. Fast methods can be useful for gathering initial data and testing basic assumptions, while slow methods can provide a more comprehensive understanding of the users and their needs. In a sense, this sounds quite simple and straightforward. Still, combining these different approaches to make timely research contributions is a challenge. For instance, one might move too quickly and be too early, or one might be too slow and seek to make a contribution when similar projects have been conducted, or when similar solutions or results have already been published, or when the interest in a topic has already started to fade away. For anyone who seeks to make timely research contributions, this is something that needs to be carefully considered.

Accordingly, one of the key takeaways that will emerge across the following chapters of this book is the importance of considering ways of *combining fast and slow approaches* to ensure timely research contributions. Another important takeaway is the role of collaboration and co-creation in interaction design research. Both fast and slow approaches often involve collaboration with users, stakeholders, designers, and interdisciplinary research teams. By involving users in the design process and working in cross-disciplinary research teams, design researchers can gain valuable insights and feedback that could inform the design of more effective and satisfying interactions.

An additional reason for why it is important to be timely in terms of research interests, approach, and contributions is related to technological developments, changes in society, and changes in the research funding landscape (which typically reflect advances in technologies and changing societal needs). For instance, there are currently several calls for research on sustainability or projects related to new technologies such as AI, Internet of Things (IoT), and cyber security – topics with clear societal relevance at this moment. At the same time, these topics are rather short term as societal needs change, and new technologies are constantly replacing other (older) technologies – just look at the technology trends over the last decade.

For sure, there are multiple reasons for digging into the temporalities at play in this research landscape, but before we move further into this topic, I should also say a few words about the overarching goals of this book.

About the Overarching Goals of This Book, before We Move on

There are several goals for this book. On an overarching level, I see it as a book that provides an overview of the different methods of design research, including both fast and slow approaches. Second, I see it as an additional goal to outline and describe the benefits and limitations of each method, and when it is most appropriate to use them.

In this book, I describe and show how fast and slow methods can be combined in an iterative design research process to gather data at different stages of the design process. I also offer practical examples, suggestions, techniques, and guidance for conducting interaction design research, including selecting the right methods, gathering and analyzing data, and presenting findings. Additionally, I also provide examples of how fast and slow methods have been used in real-world design projects to illustrate the practical application of these techniques. Finally, as a third goal, I hope this book will inspire designers to find and decide upon an appropriate and timely approach to interaction design research to create more effective, impactful designs as well as more long-lasting results.

In the following chapters, we will explore these themes in greater depth, examining the different methods and techniques used in fast and slow approaches to interaction design research. We will consider how they can be effectively combined to achieve optimal results. We also consider the challenges and ethical considerations that arise in this field and examine the latest trends and future directions in interaction design research.

As a start, the next chapter (Chapter 2) is devoted to "going fast," and it is accordingly about fast methods and approaches to interaction design research, and about how to be timely and relevant. After this chapter, I turn to ways of "going slow" in the research process (Chapter 3), including not only the slow approach to research but also a discussion on long-lasting impact.

Overall, this book follows a reflective style of writing where we move from thinking about one approach to research, and how we then shift to a different approach. Hopefully, this reflective "thesis" – "antithesis" duality and structure – will help in moving forward to a synthesis on how these different approaches can be combined to produce both timely and well-grounded research in the area of HCI/interaction design research.

CHAPTER 2

Going Fast – Being Timely and Relevant

INTRODUCTION

There are of course many reasons for fast approaches to interaction design research – not at least if we consider some of the crises and challenges we are facing right now on a global level. As our world is now facing a set of urgent and global challenges, the need for solutions is greater than ever. As formulated by the climate activist Greta Thunberg, "we are running out of time!," and accordingly, we need to act fast. To address these urgent challenges, the United Nations has identified 24 global sustainability goals that need to be addressed to create a more sustainable future for all. These goals range from ending poverty and hunger, to the oceans and life under water, to promoting sustainable cities and communities. These global challenges cover not only our natural environment and climate but also global challenges related to gender, equity, and social sustainability. And here is the thing, and some kind of a paradox. These are huge global challenges that need to be addressed collectively and globally. It is also urgent, as "we are running out of time," but at the same time, there are probably no quick fixes (just take the current much-needed transition to fossil-free fuels as one example here). Indeed, it is a call for urgent as well as long-term commitments, and it is a call where these problems scale – from particular problems, cases and cites, to the globe as a whole, and life on earth. In short, it ranges all the way from the small-scale to the globe as a whole.

DOI: 10.1201/9781003343745-2

These challenges are of course very large, and it should at the same time be noticed how interaction design research is increasingly addressing design challenges at a global scale. For instance, we notice examples of HCI and interaction design research that right now address such large-scale topics, including for instance, social justice (Fox et al., 2016; Chordia et al., 2024; Bates et al., 2018; Sum et al., 2022), (Öhlund & Wiberg, 2025) race and the Black Lives Matter movement (Schlesinger et al., 2017; Ogbonnaya-Ogburu et al., 2020; Harrington et al., 2021; Erete et al., 2021; Erete et al., 2023), challenges related to waste management, pollution, effects of technologies on the environment (Justa et al., 2024; Thomas et al., 2017 ; Rossitto et al., 2022; Dourish, 2010), and the ongoing climate changes (see e.g., DiSalvo et al., 2010, 2014; Kuznetsov et al., 2014; Knowles et al., 2018; Mencarini et al., 2023; and Soden et al., 2021, June).

These challenges are not only large, but the time frame is also crucial here. Although the challenges are huge, we need long-term solutions and we cannot postpone this – and there is no time to just contemplate. On the contrary, we must address this wide range of challenges as soon as possible. In short, there is a need for "going fast" and to develop design approaches that enable us to be fast, timely, and relevant. Indeed, we are in a hurry to develop such methods and approaches for going fast, to find solutions, and to implement and evaluate them. This can almost be framed as design-oriented "time-to-market" research, where solutions and methods need to be developed and tested quickly to be implemented as soon as possible. It is through such actions taken, right away, that we can address these global and urgent challenges.

Going fast is also about finding concrete answers and solutions. While the slow approach is about critical perspectives, and the development of ways of seeing, including theorizing and systematically moving forward, the fast approaches are very much about fixing things, about problem-solving, and about contemporary concerns (Rogers, 2012; Stephanidis et al., 2019; Rapp et al., 2022). It is as such about addressing and resolving "real world" problems. Call for empirical approaches, hands-on research (e.g., action research), and innovations are accordingly foregrounded here. For sure, empirical research is crucial in this effort, as it allows us to gather evidence and data that can inform our solutions and close the gap between research and practice. By conducting empirical and design-oriented research the idea here is that we might find ways of closing the research-practice gap and focus on immediate and pressing problems, needs, and concerns.

GOING FAST – TO BE TIMELY?

For sure, it seems like many interaction design researchers are in a hurry. "Late-breaking results" and a focus on innovation and the "new and novel" have for instance been a central concern for the CHI conference over the last couple of decades (Wiberg & Stolterman, 2014). As HCI and interaction design researchers we are eager to keep up with the pace of new technologies being developed, and we hope that we can contribute to this fast-growing landscape of technological solutions. At the same time, we are eager to follow how people make use of new technologies and how our society changes, and we are eager to understand contemporary problems and challenges. But we are not just eager and interested in our changing society, but we also have this capacity to move fast and have real-world impact – to take this ever-changing contemporary landscape as a point of departure for the design work we do. We are not just trying to keep up with the speed of technological developments, but we also seek to contribute to this evolving landscape with our designs. Now, how is this even possible? And, is "going fast" really the right approach to take here?

Well, if we assume that interaction design research can address contemporary challenges in our society, and if we also assume that we already know the methods for addressing these challenges, then we should provide the solutions as fast as we can, right? That is, to conduct interaction design research at a rapid pace, to go fast from the problem brief to the final design. In a sense, it almost becomes a game of being faster with the solutions than how the surrounding society and its related problems emerge and develop – over time.

For sure, *going fast* has its advantages. Going fast to address and resolve pressing contemporary problems is ultimately about *relevance*, and it is about providing solutions to pressing problems at a particular moment in time. As such it is a very constructive, creative, and capable approach to solving urgent problems. And of course, our world is indeed constantly changing, and interaction design research aimed at making contributions to this ever-changing world needs to be relevant – we really need these solutions *now* (just think about, e.g., the climate and environment challenges facing our planet, and how we are literary "running out of time" in terms of solutions for a more sustainable relation to our planet). In this context, the notion of "relevance" thus implies that the research needs to be conducted now to be timely. For sure, it needs to address some contemporary concerns (Rogers, 2012). Still, this window in time can be quite narrow, or even instantly disappearing – for some pressing problems it is almost too late already. In addition to this, and if we think about our world as ever-changing, then a particular design can quickly be outdated, and accordingly irrelevant if the problem it was set out to resolve has fundamentally changed.

To catch this window in time, it is common to apply a "fast" approach to interaction design research. The activities we undertake here include several methods and techniques for moving fast. For instance, we can conduct "quick-and-dirty" ethnographic studies,[1] we can arrange meetings, workshops, and brainstorming sessions, and we do rapid sketching and prototyping to get a quick understanding of the current situation and to quickly generate new ideas and drafts of solutions. Another, more overarching approach to move fast might be to structure a larger project as shorter "work packages" so that the more long-term project can be conducted in smaller, shorter, and even parallel chunks of activities.

Overall, this strategy to apply "fast approaches" suggests that relevance is a matter of catching this window in time, that is – *going fast* is the overarching research strategy to ensure the timeliness of the research. By shortening the gap in time between the identification and formulation of the research problem and the proposed solution makes it easier to target timely problems. But, as I will discuss later in this book, this also affects the possible *size* and *scope* of a research project. For sure, one cannot go for truly groundbreaking research if the time is too limited. In a too short and too limited project, there is no time for deep thinking, for exploring alternative paths, and there is no time for any larger trial and error processes. Also, if the project is too limited, it can result in the project not being relevant as it might not be able to contribute sufficiently to an important and pressing research problem. It might be short and focused, but it might then also miss the most important things.

Now, and if we dig deeper into this strategy of going fast, we need to ask a fundamental question here, i.e., – *what does it mean to do timely interaction design research?* In a sense, the word "timely" suggests that there is this window in time that needs to be targeted, and it also suggests that there is a risk of doing research that is not timely. Further, it suggests that there is a risk of doing something irrelevant – at least for the time being (and if failing to be timely). The word "timely" in this context refers to things we do "on time," at a suitable time, or ahead of time. In interaction design research this can be tricky because the field of design is also moving – at multiple paces, and as always – *research takes time*. As we move forward through our research projects there are new challenges arising, new technologies being developed, and of course, competing research projects being carried out. In short, there are many *external factors* affecting what is timely, and what is not.

There are also several *internal factors* to consider to successfully conduct timely interaction design research. A project might start with one particular research question, but this question might need to be adjusted to be relevant at a later stage. In short, the design researcher needs to consider *the time span* of a *research topic* – and how it develops over time, and he or she needs to consider the *time span* of a *research project* – and how it develops over time. It is easy to fall into a common trap here, to think that the project is too short in relation to larger ideas or ambitions. Here it is important to reflect on what needs to be done relatively fast within the given time frame of an ongoing research project, and what is to be considered as part of a slower, and more long-term goal behind the particular research project being carried out[2] at the current moment.

Interaction Design Research Techniques – For "Going Fast"

How do we do fast interaction design research? This is a fundamental question if we want to go fast. Well, our field of research has indeed been very active when it comes to the development of methods in support of fast interaction design research. No matter if you do UX design research or experimental prototyping, there are a set of interaction design research methods that allow for "quick-and-dirty" approaches to capture a current situation, methods for the rapid generation of design ideas, and methods for the rapid development of design alternatives. These methods include for instance:

- Quick studies and surveys
- Brainstorming sessions
- Workshops
- Sketching
- The making of mood boards
- Generate design concepts
- Writing short scenarios

- Creating personas

- Rapid low-fi prototyping

- Pilot tests – to get some initial data and preliminary results

I would suggest reading the textbook *Designing Interactive Systems* by Benyon (2014) for an introduction to these different methods.

On a more general level, these methods and techniques are all about moving fast. These techniques are about quickly moving from access to data, to having some data (e.g., through quick-and-dirty ethnography), or quickly moving from fragments of ideas in the head to something that can be shared, discussed, and further developed (via for instance brainstorming sessions, sketching, and the use of post-it notes as part of the design research process (see e.g., Ball et al., 2021).

Indeed, these techniques all serve the purpose of moving quickly from initial ideas and through the early stages of a design process. Could this be an aspect of the temporalities at play in interaction design research? That is, it is easy to move quickly in the beginning, and then it takes more time when going through iterations and refinements at later stages. For sure, many approaches to interaction design research advocate for these rapid techniques at the early stages of the design research process. Maybe to kick start it? Or maybe just to move from a blank sheet of paper to something that can be discussed, refined, or challenged from alternative perspectives.

In the next section, I discuss not only the approaches to going fast, but also why it might be necessary – not just for the sake of going fast, or to start somewhere, but in order to address timely concerns.

Going Fast – Is That the Approach To Take?

There are many reasons why "going fast" might not only be an *appropriate* approach but also a *necessary* approach to take. Sometimes because there is an urgent and pressing problem in practice that needs to be addressed, and sometimes because the area of solutions is being rapidly developed. For this second case, there is almost something like a "time-to-market" competition in research. If someone else has already developed a solution and published a paper about it, the race is over. Let's look at two examples that illustrate these two to some extent.

The first example that I have has to do with a crisis on a global scale – the World War II, but foremost this example is concerned with how the results from moving quickly toward solutions at a particular moment in time also led to more long-standing results (of crucial importance for the development of the modern computer). I am referring here to the cracking of the "Enigma machine" – an enciphering machine that was employed by Nazi Germany and the German military during World War II, and on the other hand, Alan Turing's work at Bletchley on "cracking the Enigma" code. In short, the Enigma machine was an enciphering machine used by the German military to send messages in an encrypted form. And for Alan Turing to crack this machine, he had to invent the Turing machine. While his work did indeed lead to the cracking of the enigma code, i.e., he made a fast and timely contribution at that time, the more long-standing results from his work are that it also led to the development of the Turing machine and to the development of a theory about this machinery – that in turn laid the foundation for the further development of AI (artificial intelligence).[3] For sure, the research at that time was very applied, and heavily directed toward an urgent problem at the time. Still, it might be the more long-term effects including advancements in the development of AI that had the more long-lasting impact.

In short, this example illustrates how research devoted to solving a particular problem, at a particular time, can have more far-reaching and more long-lasting impact – if being able to distinguish between what is needed at a particular moment, and what is needed in terms of particular solutions vs. what the more general and long-term takeaway is from the research being conducted.

The second example I would like to mention here has to do with how to make timely contributions to fast moving areas of development. For instance, if you are an interaction design researcher in an area where the pace of technological developments sets the scene and the frame for your work, then you need not only to contribute to the current existing situation, but also to keep an eye on how the field develops over a set of generations of new technologies, and look for what others are doing in this area.

For instance, if you research 4G mobile apps, then you need to be aware of the current transition toward 5G mobile networks, and maybe also look into technical research that is currently exploring what the next generation of mobile networks (6G) could offer, in terms of bandwidth, speed, security, encryption, coverage, and range. You probably also need to understand how people use 4G mobile technologies today, and how the use of such mobile apps works in their everyday lives.

Other examples here include the increasing interest right now in generative AI and LLMs – Large Language Models. All of a sudden, lots of researchers now want to be first with a published paper on these technologies, what these technologies can enable, and to also publish critical papers about the associated risks, including ethical problems with the use and development of generative AI.

If you are doing interaction design research in one such fast moving area, it is likely that you also need to adjust the pace of your research to make timely contributions. You need to understand the "now" and at the same time contribute to that particular "now" – before it has already transformed into a new state. Accordingly, such research becomes a chase for staying up to date and a chase for constantly being on the cutting edge of how practice develops. Almost to the point where we can start to talk about "time-to-market" research.

There are indeed great risks associated with this research strategy. For instance, how do you ensure that you fully understand the current situation? If it is constantly changing, and if you only have very limited time to study it, then what can you really say about it? Further, how do you do these quick rounds of research, and still make sure you are rigor enough to make a significant contribution? And how much time do you dare to spend studying the current, before moving to data analysis, and writing a paper about it before the situation changes? In a fast-paced and ever-changing landscape, it is a risk associated with stopping to observe, collect, and analyze data, draw conclusions, suggest implications, and publish the results. Indeed, focusing on the ever-changing present, and going fast is not easy, although each method and technique at a first glance looks quite straightforward and rapid, at a first glance.

Being Timely

Indeed, going fast is driven by a need to be timely – in research contexts with urgent problems, or in research contexts where the defining circumstances are changing at a rapid pace. That is to conduct research that is relevant at the time being and to arrive at results and conclusions – on time, or at least in time (before the area shifts toward other problems or relies on alternative solutions). Being timely is about that fit between what you do in your design research project, and how practice changes and develops.

Being timely is also something that can be staged. An old problem can be articulated with a timely framing, and an ongoing long-term project can be presented as being exactly about the current problem at hand. These are tactics for making design research timely. But why are these tactics needed in the first place?

Well, if we take a look at a typical interaction design research project, we can see how the design process is structured and organized to play out over a longer period in time. It might consist of the full cycle of initial user studies, concept development, prototyping, user testing, design iterations, more user testing, implementation, evaluation, another round of redesign, and so on. Of course, these activities take time. In some cases, it might take a couple of years to go through all of these activities and stages.

On an overall level, it is a predefined process that steadily steers toward some results, within a given time frame. Still, the time frame might be too long given more rapid changes in practice. Accordingly, and in many cases, the need to move quickly arises as a strategy to ensure timely results.

So how do you know when you can just follow the plan, and when it is time to change strategy and hurry up? This sense of time, and sense of how practice changes, is essential to being timely. Without this sensibility of when it is time to change the pace of the project, and without the ability to wrap things up in time – there is no way of being timely.

But being timely is not only about doing things on time – it is also about doing the right things, on time. Accordingly, we will in the next section focus on a pair of important matters, that is to be timely, and to be relevant.

Being Timely and Being Relevant – At Later Stages

Being timely and being relevant sounds like the same thing – and it maybe also sound like something that needs to happen simultaneously. However, this is not always the case. For sure, it is easy to think about these two as just two words for the same thing, as it is often the case that being timely is a necessity for being relevant. And vice versa, if you are relevant, it is likely that your contribution is also very timely. Still, there are also examples where these two are separated, and this opens up for alternative ways of thinking about just how urgent something is, and to what extent timeliness and relevance need to come together at a particular moment in time.

For instance, and as we saw in the example with the cracking of the enigma code, it was at that time considered a pressing problem, but although the solution offered at that time was relevant, the design turned out to be even more relevant at later stages in history (to the extent in this case that it has been labeled a "forerunner to the modern computer").

Indeed, interaction design research is ultimately about this interplay between fast and slow movements, about solutions to urgent and pressing problems, and about the more long-term, long-lasting impacts, or turns that define how a field develops – over time.

If we look at this particular example, we can see how "being timely" is a matter of working on a relevant (contemporary) problem and how it is about making contributions to that contemporary problem. In this case, being timely and being relevant co-exist at the same moment in time. And again, the lessons learned in that particular moment can turn out to be even more relevant – at later stages.

One can also be relevant, but be slightly too late. For instance, to contribute with a published paper with some novel results, but while waiting for the paper to appear in print some other paper with a similar case, focus, or contribution is published at an international conference, or in a related scientific journal. In fact, in fast-growing fields of research, this happens frequently.

Being relevant can also happen if some early work is picked up, reread, or reinterpreted at a later stage, where previous results – when considered in a new context suddenly help in questioning or rethinking the current moment, or serve as a vehicle for moving forward (remember the "looking back"/historical approach for moving forward that was introduced in Chapter 1).

Being Fast – A Particular Solution as a "Theory-in-Action"

If we now think about what we do when we move quickly forward in our interaction design research projects – when we have progressed at a rapid pace – we can think about this as working on particular solutions with a "theory-in-action" approach. Across the following sections I describe what this "theory-in-action" is about by using some of my own experiences from doing interaction design research projects as examples – including the development of prototypes, and presentations of how these prototypes, as manifestations of design ideas, relate to more general ideas, concept designs (Stolterman & Wiberg, 2010), or "classes-of-solutions."

A "theory-in-action" approach is about relying on an idea of how something will work, and then seeking to manifest that idea in a particular design. It is about articulating how things are expected to work, and how things are thought of as being interrelated in particular ways (and to do so on a level that is concrete enough to be articulated and manifested in a concrete design). In our previous work, we have described this as a concept-driven approach to interaction design (Stolterman & Wiberg, 2010). The idea here is to develop concepts, and particular ideas for how the design will work in a particular context, and then seek to design something that corresponds to this specific guiding idea. That is the first part of this "theory-in-action."

The second part of "theory-in-action" has to do with putting these ideas to work in the design process. As we approach our design this means that we will not only pay close attention to how the particular design is formed, but we also need to keep an eye on how it develops in relation to this particular idea, or guiding "theory."

If an interaction design researcher can do this it will result not only in a particular design, but also in quite well-articulated ideas about how the particular manifestation, i.e., the design is an example of, or corresponds to a more general idea of this design when brought into a particular use context.

This speaks to its practical contribution. But this focus on "availability management" was also important from a theoretical perspective, and here I contributed to understanding such social processes by foregrounding how availability is socially negotiated. Overall, this means that for the interaction design researcher, it is not enough to contribute with a particular design. Beyond this practical contribution, the researcher should contribute with a design concept that represents the "core idea" of the particular design proposal. But here is the kick-back. If both the particular design and the design concept are clearly formulated and expressed, it might serve as a contribution both to a particular problem and to a more general design problem.

Let me illustrate this through the use of a set of additional examples from my research. Over the last 25 years, I have designed quite many prototype systems as part of different interaction design research projects. In each project, the idea has been to design something that addresses a particular problem, but I have also sought to articulate the broader or more general design concept in question.

For instance, the RoamWare system (Wiberg, 2001) was a research prototype system designed for groups of mobile workers to support and maintain seamless ongoing communication across physical and online meetings. The particular system developed in this project did indeed work as a concrete solution to a particular problem, but the proposed technical solution also offered more general solutions that are still relevant, although this project was carried out more than 20 years ago. For instance, the dynamic session management model that this system was built around (Wiberg, 2004), including the implementation of a connection object for dynamic connections was a forerunner to today's Bluetooth stack for mobile ad hoc network connectivity. While the RoamWare system was designed for a particular group of mobile users at that time, the fundamental principle – the idea – is still relevant now, more than 20 years later.

Another example from my research is the design of an interactive system called Negotiator (Wiberg & Whittaker, 2005). This was an interactive system designed and developed to support availability management on mobile phones. It enabled lightweight negotiations of availability for mobile phone users by enabling short messages about availability to be sent as a reply to an incoming call, instead of interrupting the current activity (as would be the case if taking the phone call). Again, this was a particular design developed as a solution to a particular problem, but nowadays we can see how this principle (Figure 2.1) is now an implemented feature on most smartphones (see Figure 2.1).

The third example I have here is from the design of wall-size interactive installations at the ICEHOTEL in the far north of Sweden (Robles & Wiberg, 2010, 2011). In this project, we explored how to blend the physical and digital in the design of the icehotel, or in other words, how to "texturize" digital materials in the physical built environment. The research project had to do with the use of digital wall-size displays, and how such materials could be used in the design of an interactive atmosphere at the icehotel. On the one hand, this was a short, fast, and applied project, but on the other hand, it generated more long-lasting theoretical results (see for instance Wiberg & Robles, 2010; Wiberg, 2014; Jung et al., 2017; Wiberg, 2018). In this context, and beyond this particular project, this was one of the first attempts to conceptually and practically blend interaction design with architecture (Wiberg, 2010), and we documented the theoretical implications from this "material turn" in HCI (see e.g., Wiberg, 2010, 2013; Dalton et al., 2016).

Today, we can see how these "design theories-in-action" have played out, and how timely efforts made at some particular moments in time are still relevant, although on a different level. These design theories-in-action guided these particular research projects, and these conceptualizations are still valid. These conceptualizations are now part of the bodies of HCI/ interaction design research we have on e.g., "mobile collaboration," "availability management," and "interactive architecture." For sure, applied projects can have more long-lasting impact if the underlying ideas are articulated.

Further on, we can think about these practical projects, and how collections of such practical and particular solutions that share similar or related "theories-in-action" come together as bodies of existing work and

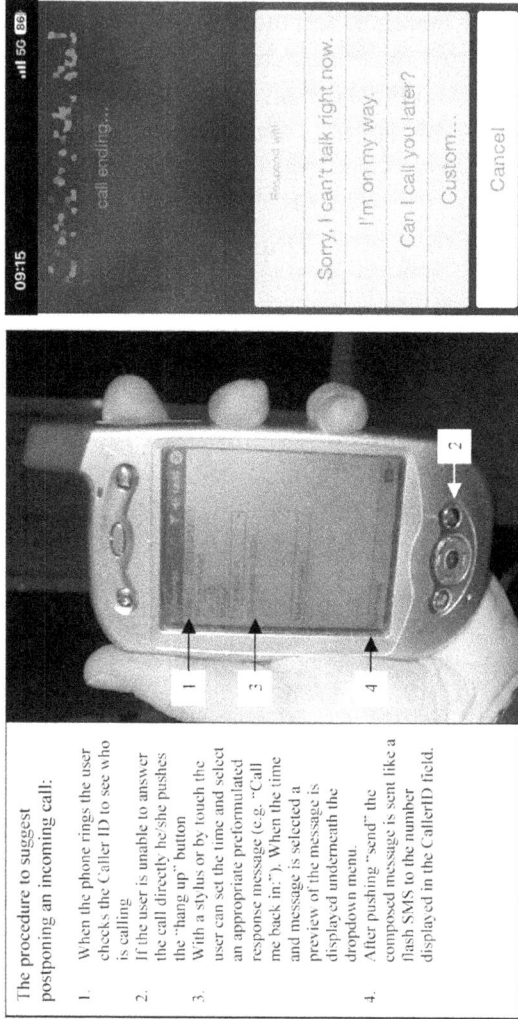

The procedure to suggest postponing an incoming call:

1. When the phone rings the user checks the Caller ID to see who is calling
2. If the user is unable to answer the call directly he/she pushes the "hang up" button
3. With a stylus or by touch the user can set the time and select an appropriate preformulated response message (e.g. "Call me back in:"). When the time and message is selected a preview of the message is displayed underneath the dropdown menu.
4. After pushing "send" the composed message is sent like a flash SMS to the number displayed in the CallerID field.

FIGURE 2.1   The negotiator system running on a Ctek smartphone (left) in 2003 and similar functionality on an iPhone (right) in 2019.

"classes-of-solutions" in these areas. For instance, the RoamWare system represents interaction design research on "mobile collaboration" as a class of solutions. The Negotiator system on the other hand belongs to a class of systems that support "availability management," and the wall-size interactive installations at the icehotel belong to a class of solutions for interactive architecture. In a paper published at the NordiCHI conference (Wiberg & Stolterman, 2014), we discussed such classes of interactive systems and how they belong to a certain idea in terms of generic design ideas and generic design solutions.

We can also see how these practical and applied projects now belong to larger research areas that have been established over the past two decades. Here we can notice e.g., the areas of Mobile Human–Computer Interaction (HCI), Computer-Mediated Communication (CMC), and Computer Supported Cooperative Work (CSCW), and most recently the emerging area of Human-Building Interaction (HBI) (see e.g., Alavi et al., 2019).

It should be said here that this transition is not about these individual projects and individual research efforts, but about a collective movement, and accordingly about multiple individual efforts, where many individual researchers have made contributions to a growing area of research, and where these individual efforts taken together form one such collective movement – a movement that over time builds a research area, and a research community where individual efforts are built upon, as well as peer-reviewed, critiqued, challenged, and rethought – over time.

While a practical interaction design research project might need to result in timely solutions, these movements toward the formation and establishment of new research areas and research communities are slower. For the individual HCI/interaction design researcher it is important to be aware of these slower movements, and an awareness of these movements is also often necessary in order to formulate timely concepts that can work as "theories-in-action." In short, this awareness works both ways – it helps in seeing in which directions the field is slowly moving, and it helps in translating these movements to concepts that can be explored at the current moment.

Being fast is accordingly not only a matter of "being in the moment" and being rapid. On the contrary, and in order to work fast, the HCI/interaction design researcher needs to be anchored in the history of the field, so that he or she can be informed and inspired by what has already been done, and be capable of seeing work that is still lacking, or be in the capacity of being future-oriented – to be capable of imagining new approaches to explore, or formulate new perspectives that allow for new ways of seeing, approaching, challenging, and designing. In a sense, following these movements in a field, and contributing to such movements is like contributing to a slow conversation in a field. A slow conversation that plays out through the projects conducted, and through the papers published about these projects – where links are established to the ongoing conversation in the field not only by the contributions offered, but also through references and presentation of related work.

Despite these slow conversations, there are also good reasons why HCI and interaction design researchers need to be fast, and move quickly toward solutions. Typically, it has to do with the fact that there are always external stakeholders surrounding design research projects. In the next section, I therefore ask – timely and relevant design research, for whom?

Being Timely and Relevant – But for Whom?

It is frequently described how design is always done in relation to different stakeholders – no matter if the stakeholder is a client, an owner, someone funding the project, or the intended receiver, consumer, or the end user of the design.

Design is always directed toward someone, and being timely is accordingly also a question about who is listening, and who is present when you present your work.

Of course, being dependent, and acting in relation to external stakeholders puts pressure on the research process. The proposed design needs to live up to the expectations of these stakeholders, and the proposed solution needs to be developed and delivered on time. Accordingly, design becomes an act of meeting external expectations, about meeting deadlines – and quite frequently this becomes an act of moving fast to make this happen – to deliver to someone, and deliver on time.

When being timely and relevant becomes a matter of being timely and relevant concerning these stakeholders the design process can easily become a stakeholder game. It is often the case that many stakeholders want the same thing, but it can also be the case that there are conflicting interests.[4] Such conflicting interests can affect the design process in terms of what is prioritized, in which order things are addressed, resolved, and delivered, what the end result is, and how it is articulated.

There are also stakeholders operating at many different levels. On a practical level, it might be the client, the end user, and the people funding the project, but on a more overarching level we should also consider how societal movements (i.e., today's foregrounding of sustainability) or political goals put pressure on what is researched and designed, and how fast solutions are expected to be delivered (just think about how our society rearranged the processes for producing a vaccine against COVID-19 so that instead of a research and approach period of 4–10 years, the vaccine was now developed, tested, and made available to the public through mass vaccination in less than a year after the outbreak of this global pandemic). Indeed, external pressure from powerful stakeholders can come in many different forms, and it can come from different stakeholders. With this, it might also imply shorter time frames, and come with expectations for moving fast toward expected solutions and results.

Another set of stakeholders are the partners with whom design research projects are conducted. It is nowadays quite common to work not only with the intended end users as clients (the typical setup in many participatory design project), but to also conduct design research in large-scale research-industry consortiums. In these formations, it is common to talk about, e.g., "engaged scholarship" (Glass & Fitzgerald, 2010) or "action research" (Cohen et al., 2017; Avison et al., 1999; Checkland & Scholes, 1999) where a central idea is that the design researchers serve as a long-term and active partner in these larger project consortiums. Of course, being part of one such large-scale consortium also comes with expectations on not only the expected outcomes of the research process, but also expectations about *when* certain results and outcomes can be expected.

Given this it is possible to see at least five sets of stakeholders with different needs including the following:

1) Client needs

2) Pressing societal needs

3) Political goals (e.g., "fossil-free society," EU goals for 2030, sustainability, etc.)

4) Funding agencies (typically aligned with political goals), formulated calls for research

5) Industry collaborations (access to data, relevance, "engaged scholarship," etc.)

Being timely is not only a matter of presenting results to different stakeholders and meeting external expectations. It is also a matter of how "now" is understood and described. In the next section I describe how this becomes almost a philosophical matter, but still a matter with very practical implications for what it means to do timely and relevant interaction design research.

Being Timely and Relevant – Now

It sounds quite easy to address urgent problems, to live up to stakeholders' expectations, and to deliver results on time in order to be timely. But is it really that simple? Well, it is not. Design research is situated right in between design and research. It is about designing – in relation to user needs, more-than-human needs, stakeholders' needs and expectations, and so on, but these efforts also need to contribute to an existing body of published research for it to be timely design research. Accordingly, design research becomes an act of simultaneously making things happen – now, while making sure that it adds to an existing body of published research. And, in addition to this, design research needs to make contributions that add to a preferable future. As such, it is also fundamentally value-driven.

This means that being timely becomes a matter of not just understanding different stakeholders' needs and expectations, but also a matter of understanding "the now" through an understanding of the past, and understanding how alternative preferable futures can be made possible by reimagining the now – through design. It is an act of addressing the now, and it is an act of transition, from the present, the current now, to a future state.

In this transition, it is important to ask critical questions regarding "whose now" the design research serves. That is, to think critically about which stakeholders are served by, or privileged by, the design research (and who might be marginalized or even rendered invisible). At the current moment, there is a turn toward "more-than-human" HCI/interaction design research. The motivation for this turn is simple. For decades HCI/interaction design research has foregrounded the importance of being "human-centered" in the approach. In a sense, the human-centered approach has served as a guarantor for good design – if the design serves human needs, then it is good, right? But the problem with this is that humans are in a privileged position, and there are other life forms on this planet. So why not design not only for humans, but for more-than-human needs?

Further, it is also about critical thinking on "which now" and "which futures" the design research serves and enables. With only one particular stakeholder in mind, it can be misleading to think that there is only one path forward, and it might be hard to see alternatives and tradeoff costs. However, there might in fact be many alternatives, and for the design research process it can be helpful to examine existing and potential stakeholders to identify such alternative paths and alternative visions of future states.

Here we should also notice how "the now" is something that is co-constructed (by actors defining how to understand the "now," in articulations of what is foregrounded in descriptions of "the now" and in relation to what technologies we foreground as new, relevant, interesting, etc). As an example, we can notice how artificial intelligence (AI) is at the current moment foregrounded as the latest "now technology," and also how we have seen this before. For instance, how 3G was described as a revolutionary mobile technology at that time, how "Web 2.0" was described as a game changer for the Internet, and how the Internet of Things (IoT) was described as a paradigm shift of computing that would enable any physical object to have computational capabilities and become accessible, anywhere over the network.

If we subscribe to this idea of how "the now" might be about "multiple nows," and how these nows are co-constructed by multiple actors and stakeholders, we can think about this along the ideas formulated by Wakkary (2021). According to him (see Wakkary's notion of "multiplicities"), design research is about transitional acts, where the design researcher should focus less on what is and more on multiplicities of what could be (future making as an act of destabilizing, and accordingly being about many possible futures, multiple futures, for the many). Along these lines, a number of researchers are now exploring the pluriverse, in terms of searching for alternative futures, and the making of alternative worlds (see e.g., the work by Escobar, 2011 ; Querejazu, 2016; Escobar, 2018).

To reflect on the now, possible future nows (with "nows" in plural form), and which stakeholders might be served in such transitions becomes on the one hand something close to a philosophical issue (including the questions of, e.g., "what constitutes the current moment" and "who forms the now and the future"), but on the other hand, this can be thought of from a very practical perspective. In going back to the previous section, design is a relational activity – an activity carried out in relation to an external stakeholder (e.g., in relation to a client, a user, an owner, global needs), and accordingly, being timely, and to contribute to a particular now then simultaneously becomes a practical matter of making sure that the design corresponds to the expectations and demands expressed by these stakeholders.

We can take UX – User Experience design as an example here – User experience is related to user needs, and to what extent these users can use the design (usability). It is also related as to the users' expectations of how something should work. Over time, a solution that worked say 10 years ago on a computer could at that time have been experienced as functional, fast, "high-tech," and smooth, but now, 10 years later the same solution can be experienced as slow, "old fashioned," and clumsy. This is simply because we now have other expectations on technology, and we have other experiences from similar but faster solutions, etc. Accordingly, to be timely is a relational concern – both related to external actors, but also related to expectations at the time being. If the user expects a certain solution, the design is only timely if it aligns with these expectations.

This brings us to the issue of making timely and novel contributions. Central to design research is that the solutions should be new and novel. It is not just enough to solve the problem, but the solution should also be new and unique. This is radically different from other research fields where the focus is on understanding and ultimately solving a problem. In traditional fields of research, progress is made when a research problem is solved, so that the field can move on to address the next research problem. For the traditional fields of research, especially in the natural sciences, it is also good if the research can be repeated, that is "repeatability" ensures the method and approach, and it opens up the possibility of repeating the research to further confirm that the approach leads to the same results and outcomes. Accordingly, research then becomes this act and process of stabilizing methods and precising the results – over time.

In design research, the fundamental logic is somewhat different. It is of course still important to address timely research problems, but a contribution is typically made through a novel approach, or through the presentation of a novel solution. Instead of "repeatability" it is considered important to explore new grounds, concepts, and techniques, to present new designs, new solutions, and new approaches – and through this process of explorations, the idea is that we expand our field in new directions.

Related to this we can see how the notion of a clear "research front" and the metaphor of "pushing the research front" is somewhat problematic in design research. This metaphor of a research "front" gives the impression of ONE front, and one front LINE, and that this line can be pushed forward through research efforts. Of course, if you work in a traditional area of research you can probably see how someone managed to solve a particular research problem, and you can repeat the process to inspect and verify it, and given that it becomes an established technique you can move on to address the next research problem. In short, it is assumed that there is a clear "line" in the "now" and that this line, this research front line, can be addressed and pushed forward.

But in design research, we do not really have this clear systematic and cumulative approach. Not because we do not want to, but because we focus on particularities, multiplicities, and explorations. In short, we focus on ultimate particulars. This leads us to focus on what is unique in each design process, it leads us to think about many alternatives (and we do this when we ideate, when we brainstorm, when we sketch, and when we iterate). We are not interested in the one and only solution, but in thinking about what possible futures we can imagine – to rethink the current from the viewpoint of multiple possible futures.

Again, this brings us back to more philosophical matters here. According to Nelson and Stolterman (2014) we can think about the "now" and how it is related to "the future" in terms of "the now" (the present), and the "not yet existing now" (the future). In short, design then becomes an approach for moving toward "the not yet existing," and at the same time, this approach holds a promise of making this existing, a promise of making the new now. For the design researcher, the issue of being timely then becomes a concern for being timely in the creation of this new now.

This strives in design toward "the not yet existing" also tell us something more about the fundamental nature of design, that is, we are fundamentally future-oriented in our approach to "the now," and accordingly, to be timely becomes an act of thinking ahead (of now), and an act of being future-oriented.

While the traditional sciences might be occupied with related work to understand the current situation and research landscape, and to identify the current research gaps, design research still care about related work, but not only to understand the present situation, and to identify the gaps, but also to see what approaches that has been proposed for moving forward, and what designs that has been proposed in this orientation toward the future.

So how can we navigate toward the future? We need visions, for sure. But we also need to formulate ideas about the future. This is where prospects play a crucial role.

Being Timely, but Future-Oriented – On the Importance of Prospects

Being timely can be seen as a limiting perspective if seen only as an act of adding to something that is now, in terms of the current moment. Not at least if thinking about "the now" as something that is in constant transformation, and accordingly something that is also always gradually disappearing as a result of this transformation from the current now, to the future now. Still, this urge to contribute to the current now, to the existing situation, has been a dominant perspective, and accordingly, it comes without surprise that many design researchers feel they are in a hurry to make their contributions fast, while there is still time, i.e, before "the current now" changes to something new.

However, as we saw in the previous section there are alternatives to this approach. If viewing design as an approach to "the not yet existing" and if thinking about design as an act of future making, and as such an activity concerned with the making of new nows, then this future-oriented approach opens up for alternatives, away from being in a hurry, and instead in favor for activities that works toward the establishment of new future grounds.

What follows from this is that we can think about "being timely" as a matter related to some ontological concerns, that is, "how things are, right now" (and how then to contribute to this existing "now"), whereas "being future-oriented" becomes an issue less concerned with "what is" but more about "how things could be." That is, being future-oriented is a concern for *prospects* rather than *ontologies*.

So, what then is a prospect? And how is it different from an ontology? Well, it is commonly described how ontology is an area of philosophy that studies aspects of "what is" from the viewpoint of e.g., existence, being, becoming, and reality. It includes the questions of existence on the most fundamental level. It is occupied with matters of existence rather than with "what could" exist. On the contrary, prospects are about *the not yet existing* and about "what could be." Merriam-Webster dictionary defines prospects through a set of interrelated notions including *anticipation, vision, lookout, possibility*, and *scene*. From the viewpoint of anticipation, the presentation of a prospect is "an act of looking forward." It suggests that a prospect brings an extensive view to the current situation, but allows for extending the now into the future (implied by "looking forward"). On a related note, and if prospects are less about "what is," but more about this act of looking forward, toward "how things could be" it is also about possibilities, and as such it can even be about things that are awaited, expected, or hoped for. A prospect can also be in the form of a vision, a picture of something to come. Again, it has to do with a future state, and it has less to do with the current situation. A prospect can also be articulated in the form of a lookout, as something that suggests an extensive view. Finally, a prospect can take the shape of a scene, a vision that helps to extend the view in how it illustrates how things could be, and how things can be thought of and interrelated.

At the core of a prospect is the notion of "could" and "what could be," that is, the possibility, rather than "what is" (which is again more of an ontological concern).

For the design researcher, the overarching implications from this are two things. First, to be future-oriented in the research approach means to work toward the formulation of such prospects. And second, to be timely has to do with either making contributions to an established prospect, as to make that prospect real (that is, to make "the not yet existing" real), or to conduct design research on the formulation and exploration of prospects that articulates new possibilities, or foreground things that are needed, awaited or expected.

In the area of HCI (human–computer interaction), we can see how we have already been through three such large-scale prospects, typically described as the three waves of HCI.[5] And while the description of the first two waves was to a large extent about providing some perspective on how the field had developed, it is fairly easy to see how the prospect of third wave HCI has guided lots of design research to explore digital design in people's everyday lives. Further, and along the lines of being future-oriented, we can now start to see examples of HCI researchers who are starting to formulate prospects for "the fourth wave" of HCI[6] and how this is then further explored through research programs and design research projects.

Related to this we can think about "being fast" as an approach to *within-prospect* design research, whereas the formulation and formation of new prospects usually takes time. Still, we both need the prospects, and the design research that makes these prospects part of our reality. This is indeed a delicate interplay of temporalities in design research, and in the next section I present an overview of a number of such prospects that have guided people to conduct timely design research in the area of HCI.

On Grounded Approaches – To Timely Challenges

What we saw in the previous section was how a prospect can provide a frame and a vision for moving forward. For any design researcher who seeks to be timely, this is of great importance. That is, we are dependent on the existence or the formulation of these prospects in our efforts to make timely contributions.

Still, the act of formulating new prospects takes time. In the area of HCI, which is an area that has strived for the integration of humans and digital technologies (not at least with "usability" as a key concern), we can notice how the formulation of new prospects is not only about understanding and articulating the next steps forward in terms of new technologies, but also a matter of understanding the use of these technologies, a matter of understanding how people and society is changing, and a matter of how digital technologies can be part of such changes. This is further related to understanding social structures, social dynamics, values, ethics, history, culture, and politics. Indeed, the formulation and exploration of such prospects take time.

Further on, we can quite easily see when such prospects are too utopian and technology-driven (and sometimes we refer to such accounts as promoting a worldview of technological determinism). On the other hand we sometimes see proposals that are clear about the value grounds, and proposals full of hopes for a better, more just, or more ethical future, but where the prospect lacks details about the enabling technologies, or identified barriers for development.

We have in the area of HCI worked our way across a number of powerful prospects that has managed to strike a balance between a clear vision of a future technology and what it would mean for people using it. If I should name just a few of these, I can at least point at six such prospects that have been formulated and explored over the last 25 years in our area of research including 1) *groupware*, 2) *mobile computing*, 3) *ubiquitous computing*, 4) *tangible computing*, 5) *social media*, and most recently 6) *AI – artificial intelligence*.

Now, if we quickly examine these six prospects, we can see how each one of these prospects included a particular strand of digital technologies, and a vision for how the technologies would be used and utilized. In addition, and on a more general level, each prospect is both future-oriented, in serving as a *vision* for moving forward, as well as it provides a *scene*, or a frame for people to do research within. These prospects have also been developed in a timely way so that each prospect also speaks to possibilities that are within reach (due to technological advancements, trends, or spread in terms of use and adoption), and as such serves as frames of articulations that foreground things that are needed, awaited, or expected.

Prospect 1 – For instance, if we take the first example of Groupwares, we can see how this prospect was formed around emerging technologies for online communication (including better computers, bandwidth in computer networks, etc.), but also formed around ideas of "a global village" where everyone is connected, and ways of reimaging work in a computational moment (thus the focus on, e.g., CSCW – Computer Supported Cooperative Work that grow out as a particular research area around the development of groupware in the 1990s.

Prospect 2 – The second example, with the prospect of "Mobile computing," was driven both by a vision of providing an alternative to "stationary work," and not at least stationary or location-bound office work which had been in focus for CSCW. It was a prospect that suggested an alternative to desktop computing, and it was driven by a vision of "anytime, anywhere" computing, mobility, connectivity, and work. It was a prospect that wanted to set people free, and challenge geographic restrictions, and at the same time it was also a prospect that grew out from technical advancements made, including at that time the development of handheld computers, the first smartphone devices, and at that time, the introduction of 3G mobile networking that allowed for real-time, online mobile services (nowadays referred to as mobile apps).

Prospect 3 – With mobile computing in place, we can see how a third prospect was formulated around "ubiquitous computing." This prospect was not just about mobile computing, but computing enabled and accessible everywhere, and computational power embedded in literary every thing – in everything. We can see how this prospect served as a frame for the development of online digital services and cloud computing (along with the idea that it did no longer matter on which machine the computing happened, it could happen anywhere "in the cloud"), and how this prospect worked as a vehicle for generating further ideas about computing in alternative forms and manifestations – as "anywhere computing" (in the development of cloud services) and computing in "every thing" (as sensors and embedded systems).

Prospect 4 – These fundamental ideas about computing can also be seen in the related prospect of "tangible computing." While the prospect of ubiquitous computing was about computing "every-where" and in "any form," the prospect of tangible computing was more specific – to the point that it wanted to serve as an alternative vision for human–computer interaction, a vision of making comput-ing possible and accessible not only in digital form – in bits, but also in physical and material form – in atoms. A vision about the design and use of Tangible User Interfaces (TUIs) that would allow physi-cal things to have computational qualities, and a vision of how to enable people to interact with computational matters through physi-cal objects. While this was very visionary in the early days of this prospect, we can now see how it has been established both in prac-tice (with a growing interest in the development of IoT – Internet of Things technologies) and in theory (with materiality being estab-lished as a theoretical ground for such systems that operates across and integrates physical and digital materials).

Prospect 5 – In addition to these *prospects*, we can see how "Social media" has served as a guiding prospect for thinking about the inter-net as a platform for social interaction. In the early days of social media, it was discussed under notions such as "new media" and as a way to democratize media. We have also seen how it has sparked a number of streams of research, for instance on online privacy and integrity, on user-generated content, and maybe most recently on what has been labeled "the sharing economy." This prospect has also led to the development and establishment of a whole vocabulary con-cerned with what people do on these platforms, including the nowa-days common notions "post, like, share, follow, etc.," and the related notions of "wall, channel, comments, followers, and hashtags." It has also led to research on emerging phenomena on these social net-works, including studies of things that "goes viral," the emergence of "influencers," and how businesses can work with "online presence" in new ways.

Prospect 6 – Finally, the sixth and last prospect I would like to mention here might also maybe the most recent one, although it might be the third wave of a prospect that was initially formulated in the 1960s (Xu, 2019). What I am thinking about here is the prospect of AI – artificial intelligence, or maybe more specifically about the third wave of applied AI with a focus on exploring ways of applying and using AI for different purposes, and in different social contexts. As formulated by Holmqvist (2017) , AI is now available in forms that make it applicable and suitable for integration in systems and user interfaces. This expands this prospect from the research labs to the contexts of people's everyday lives, to the whole society. Accordingly, and as AI is spread on a societal level, we can see how this prospect opens up for research on, e.g., AI transparency, AI and ethics, power, and ways of integrating AI in decision-making processes.

There are of course lots and lots of additional things that should be mentioned about each of these prospects, and I can for sure list additional prospects here. Still, the important thing here is not to provide a complete overview, but instead illustrate how powerful and useful prospects can be for design-oriented studies in our field. Indeed, the articulation of strong or guiding prospects enables people to conduct design research on timely topics, and if people understand the prospect, it is easier to see how individual pieces of research can serve as important contributions in the realization of a larger prospect – no matter if it is about new ways of working with mobile technologies, or if it's about designerly approaches to a more just society.

Over the last 25 years, we have seen how powerful these prospects have been. Each of these prospects has provided a ground and a frame for design researchers to work on. These prospects have also provided opportunities to be forward-looking (anticipation), and these prospects have served as visions for things to come, and as a reference point for imagining and working toward alternative futures. These prospects have also served as door openers to new possibilities. This illustrates how prospects can be both useful and powerful.

In short, *prospects* need to go fast, but of course, the formulation of clear and powerful new prospects, on the other hand, takes time. So how do we formulate such prospects? In this book, I suggest that this question points to the need for a deeper understanding of the importance of slow approaches to design research. Accordingly, the next chapter is devoted to how design research can be anchored in the growing body of research on what has recently been called "slow science." Here I relate design research to this growing perspective and I discuss the notion of "prospects" in relation to Kuhn's notion of "paradigms" (Kuhn (1970). I also go into detail about the importance of basic research in design, and I will discuss a set of thought styles as ways of thinking, and ways of exploring new theoretical grounds for design research.

SUMMARIZING THE FAST APPROACH TO DESIGN RESEARCH

In wrapping up this chapter, I would like to provide an overview of the fast approach to design research, including an overview of the most common methodological approaches, some of its advantages, and some recommendations for when a fast approach might be the preferred pace for moving forward.

As a recap, the fast approach to design research is sometimes labeled as "quick-and-dirty" research, which might involve collecting data quickly and efficiently through methods such as rapid prototyping (see e.g., Upcraft & Fletcher, 2003 ; Tripp & Bichelmeyer, 1990; Wilson & Rosenberg, 1988 ; Yan & Gu, 1996; Kamrani & Nasr, 2010; Jones & Richey, 2000), user interviews, ethnography (Hughes et al., 1995), focused ethnography (Knoblauch, 2005), short-term ethnography (Pink & Morgan, 2013), quick ethnography (Handwerker, 2001), rapid ethnography (Millen, 2000), and even "quick-and-dirty" ethnography (Valentin et al., 2012; and Vindrola-Padros, C. & Vindrola-Padros, B., 2018), and surveys. It might include Wizard of Oz studies, short ethnographic studies, workshops, focus groups, and the development and use of low-fi prototypes. These fast approaches to design research refer to methods that allow designers to quickly gather insights and information about users and their needs, preferences, and behavior. These approaches typically focus on identifying key issues and opportunities as efficiently as possible, rather than collecting exhaustive data. The fast approaches are indeed geared toward moving faster forward rather than stopping, reflecting, revisiting, rethinking, finding connections, unveiling the hidden logic, and theorizing. In that sense, the fast approaches are about doing rather than about understanding (although of course, one needs to have a good understanding to take the right actions). More about this toward the end of this book.

Some common fast approaches to design research include both empirical methods as well as methods for quick prototyping and testing of ideas. For instance, user interviews involve speaking with a small number of users over 1–2 hours to gather insights about their activities, experiences, needs, and behaviors. These interviews can be conducted in person or remotely, and can be structured or unstructured, sometimes referred to as semi-structured, depending on the goals of the research.

Another option is user surveys. This is a quick and cost-effective way to gather data from a large number of users. Surveys can be administered online or in person, and can include questions about users' experiences, preferences, and behaviors. The advantages of using user surveys over interviews are related to getting an overview of a user group, and collecting opinions that can be easily rated on different scales. Surveys can of course include both multiple-choice questions and free text fields, but if one wants to understand underlying motives, feelings, etc. then in-depth interviews should be considered instead of a broader survey.

Having collected some user data, it is now time to move toward the first design. For this phase, there are also a set of fast approaches available. For instance, rapid prototyping and iterative testing are two examples of fast approaches to design research that involve creating and testing prototypes of a future product or service in rapid succession. The goal of these methods is to quickly gather data and feedback on a design, in order to identify and address problems or opportunities as early as possible in the design process.

Here, rapid prototyping (sometimes referred to as "low-fi" prototyping (see for instance, Sefelin et al., 2003; Walker, 2002; De Sá & Carriço, 2006; Rudd et al., 1996), involves creating a simplified version of a product or service, often using low-fidelity materials such as paper or cardboard. These prototypes can be quickly and inexpensively produced, and are used to test basic assumptions and gather initial feedback from users. This can be combined with quick cycles of iterative testing which involves testing the prototype with users and gathering feedback, then making changes to the design based on that feedback. This process is typically repeated until the prototype meets the needs of the users. Both rapid prototyping and iterative testing are useful for gathering data quickly and efficiently (Jones & Richey, 2000; Camburn et al., 2017; Neeley et al., 2013; Arnowitz et al., 2010), and can help designers to identify and solve problems early in the design process. These methods are particularly useful for testing early stage concepts and ideas, and can help designers to save time and resources by identifying and addressing problems before they become more difficult to fix.

Having been through the first design iterations there might be a need for some rapid feedback on these design ideas. If instant feedback on some design ideas, prototypes, etc. is needed then there are again a number of fast approaches available. This ranges from simple workshop formats to more structured focus group sessions and even more formal usability testing. In this context, usability testing involves observing users as they interact with a digital product or service to identify any issues or difficulties they may experience. This can be done through in-person testing or remote testing using tools such as screen-sharing software, one-way mirrors, video cameras, etc. Focus groups (Wilkinson, 1998); Bertrand et al., 1992), on the other hand, involve bringing a small group of users together to discuss a product or service and provide feedback. These focus groups can be also conducted either in person or remotely, and are usually facilitated by a moderator.

These fast approaches to design research can be useful for quickly gathering insights and identifying key issues, but it's important to keep in mind that they may not provide a comprehensive understanding of user needs and behaviors. In some cases, slower, and more in-depth research methods may be necessary to fully understand user needs and design solutions that meet them – more about this in Chapter 9 where I discuss ways of combining the fast and slow approaches.

Overall, it may be necessary to adapt the methods and techniques used in the fast approach to ensure that they are appropriate for the overarching research question and context, and on the other hand, to spend time on the crafting of the overarching prospect – the vision or formulated path for moving forward. In short, to ask fundamental questions about the aim, the goal, and the more long-term vision that motivates the research project in the first place. Then, when knowing more about the ultimate aim one can seek to adjust the planning and the timing of different research activities to ensure timely contributions.

As I will set out to illustrate in the next chapter (Chapter 3), this dwelling, these thoughtful approaches to the exploration of new grounds, and the formulation of interesting and well-grounded prospects are essential steps to take to conduct timely design research, and to make timely research contributions. Accordingly, the next chapter will provide one such additional perspective – the perspective of going slow. With these two perspectives in place, including ways of going fast, and ways of going slow we can start to understand the interplay of fast and slow in design research, and ultimately the rhythms of design and the paces of research. As I suggest in this book, this dynamics is crucial to understand to do timely research and to make timely research contributions.

NOTES

1. For a discussion on the use of ethnographic methods in design-oriented research see e.g., the work by Nardi (1997). Nardi (1997). The use of ethnographic methods in design and evaluation. In *Handbook of human-computer interaction* (pp. 361–366). North-Holland.

2. In Chapter 3, I expand on these ideas, including "slow science," the development of research programs, the importance of striving for change through long-lasting impact, and the pace of theorizing in design research.

3. See the classic paper "Computing Machinery and Intelligence," a seminal paper written by professor Alan Turing on the topic of artificial intelligence. Turing (2009). *Computing machinery and intelligence* (pp. 23–65). Springer Netherlands.

4. A wonderful example of a methodology that was explicitly developed to capture different stakeholders' interests, and make visible potential conflicting interests can be found in SSM – Soft Systems Methodology, an approach originally developed by Peter Checkland as part of his work on systems thinking.

5. For an in-depth presentation and discussion on these "three waves of HCI" see the classic paper *"When second wave HCI meets third wave challenges"* by Susanne Bødker (2006).

6. See e.g., *Christopher Frauenberger. 2019. Entanglement HCI The Next Wave? ACM ToCHI – Transactions of Computer-Human Interaction 27, 1, Article 2 (January 2020), 27 pages.* DOI:https://doi.org/10.1145/3364998

CHAPTER 3

Going Slow – And Making Long-Lasting Impact

INTRODUCTION

Across a wide range of academic disciplines, the idea of "slow science" is now spreading (see e.g.,. Picho et al., 2016; Frith, 2020). For sure, the idea of "slow ideas" (Gawande, 2013) and ways of approaching the world slowly but steadily is a growing approach, and it rhymes well with the idea of science as being about cumulative knowledge, where we slowly accumulate new knowledge and add to an existing body of research (Picho et al., 2016).

In this chapter, the idea behind the notion of the "slow professor" (Berg & Seeber, 2016) is introduced, along with the concept of "wicked problems" and the need for thoughtful and critical approaches to address such problems. The concept of wicked problems was developed in the planning literature (Rittel & Webber, 1973) to describe emerging policy problems that did not correspond neatly to the conventional models of policy analysis used at the time. This notion has then been further developed and explored by a great number of researchers: see for instance, the work by Skaburskis (2008), Lönngren and Van Poeck (2021), and Buchanan (1992).

In addition to this, and along the philosophy of "slow science," this chapter highlights the importance of "staying with the trouble" (Haraway, 2016) and to consider fundamental or even eternal research questions,

 DOI: 10.1201/9781003343745-3

rather than seeking quick solutions to small-scale problems. Here, eternal research problems are those that are always important, for instance, how to develop a more sustainable society, or how to address problems related to inequalities in society. It can also be about the fundamental assumptions we have in our field. For instance, the idea of being human-centered in interaction design and then returning to this from a multitude of perspectives. For instance, should we be only human-centered? Then what about other species? And what about the environmental footprint that follows from a human-centered approach? Or to constantly come back to the central question "What is interaction?" in interaction design research. For such questions, we might not find a complete or final answer and then consider it solved. Instead, it forms the basis for a field of research, where every new study can add to our understanding of such cornerstones of a research field: in short, to avoid "solutionism," to avoid the "quick-and-dirty" approach to research, and instead move slowly and steadily forward.

The slow approach to research is typically described as being theory-driven and based on the cumulative research tradition (see for instance Owens, 2013a; Lutz, 2012; Alleva, 2006). In HCI this has been formulated as a concept-driven approach (Stolterman & Wiberg, 2010), or as an approach that works with and develops "strong concepts" (Höök & Löwgren, 2012).

It is also common to connect slow approaches to fundamental studies, including cumulative and basic research (see for instance Salo & Heikkinen, 2018; Ulmer, 2017; Carp, 2011; Owens, 2013b; Frith, 2020). Examples of slow approaches include literature reviews, slow ontologies (Ulmer, 2017), historical studies and perspectives, slow ethnography (Grandia, 2015), ethnoarchaeology (see for instance Cunningham & MacEachern, 2016; Marila, 2019), and so-called close readings. Here I would like to emphasize the importance of taking time to understand complex matters, to develop and define concepts, vocabularies, and theoretical frameworks, the importance of longitudinal studies, and why it is important to not just address obvious research problems but to also follow slow changes, adaptations, mutations, transitions, as well as resistance to change and aspects of stability over time. Overall, this chapter has an explicit focus on the value of going slow to thoroughly understand and address complex matters and issues of depth, scale, and slow changes.

The "going slow" approach to design research allows for a process that can foster a deeper understanding of processes, change, transition, emergence, and other unfolding phenomena. Further on, it allows for reflection, critique, and ways of rethinking what we think we know and to rethink the things we take for granted. While this process takes more time – by definition – the slow approach allows for more thorough and rigorous research, and as I suggest in this chapter, the slow approach can ultimately lead to more fundamental, and game-changing, designs, insights, theories, implications, and recommendations.

On a general level, the concept of "slow science" challenges the traditional notion of fast-paced, data-driven, and solution-oriented design research. Instead of pushing one such "quick fix" approach to research, the slow approach advocates for a more thoughtful and reflective method. It emphasizes how it takes time to understand, question, and rethink complex issues, as well as to consider even eternal or at least very fundamental research problems and research questions. This approach is particularly relevant in the field of interaction design, where designers often face "wicked problems" (Buchanan, 1992) that require critical, creative, and speculative thinking to address and rethink the things we so easily take for granted. Just consider the technology-oriented approach prevalent in design-oriented HCI. Is a new piece of digital technology always the best solution to an identified user need? And should the proposed solution always be built on the latest tech trend? (Right now that would be solutions based on generative AI and LLMs – Large Language Models.)

Another key aspect of the slow approach to interaction design research is that it is also highly collaborative – although operating along a different logic than the fast approach. While the fast approach foregrounds collaboration in the forms of research projects, research centers, groups or teams, or activities you might undertake as a part of the research process (including workshops, brainstorming, rapid prototyping, etc.), the slow research approach is based on another level of collaboration tightly coupled to the idea of a cumulative tradition – within a community and a given paradigm, a collective movement where we build new theories and generate new knowledge upon the work of others. It might involve long periods of individual contributions, but it is oriented toward a shared body of knowledge. In this context, this notion of "shared" means that it is collectively shared within a research community. In short, we succeed when we are "standing on the shoulders of giants." This work involves literature reviews and historical perspectives to understand the existing body of knowledge, as well as close readings to dig deeper and pay attention to nitty-gritty details and meanings embedded in what at first glance might seem obvious, simple, and straightforward. Of course, it also includes higher academic seminars, debates, and peer reviews where we discuss – over time – and critically examine new ideas presented to the research community.

Longitudinal studies (White & Arzi, 2005) and other forms of empirical research can also provide valuable insights into complex phenomena and how they change and evolve – over time. Going slow, as I suggest in this chapter, allows for a deeper understanding of processes, change, transition, emergence, and other unfolding phenomena. While it may take more time, the slow approach allows for more thorough and rigorous research and can ultimately lead to more groundbreaking solutions. In short, by embracing the concept of "going slow," researchers and designers in the field of interaction design can, if playing the game in the long run, create more thoughtful, theoretical, and groundbreaking research. For sure, to break existing grounds, and to propose such new grounds – takes time.

ON THE IDEA AND PHILOSOPHY OF "SLOW SCHOLARSHIP"

At a time when research is increasingly characterized by short-term projects, clear goals, well-structured work packages, deliverables, and rapid results, the concept of *slow scholarship* stands as a fundamentally different model – a call to embrace ways of doing research that builds on values, patience, contemplation, and depth. Slow science, slow scholarship, and the notion of "the slow professor" (Berg & Seeber, 2016) – are all variations of this shared approach to research that seeks to redefine the tempo of academic inquiry, shifting the focus from speed to significance, from project management and structures to insights and reflections. At its core, slow scholarship centers around a commitment to explorative or systematic research to either seek new directions or to systematically move forward. In this context, the systematic approach refers to a deliberate, methodical approach that values quality over quantity, depth over speed, and thoughts over processing of data. Drawing inspiration from Thomas Kuhn's notion of cumulative research and paradigms (Kuhn, 1970), the slow research approach recognizes the importance of building upon the foundational work of predecessors, of "standing on the shoulders of giants" (classic words formulated by Isaac Newton when he said that his success had been built on the achievements of others: *'If I have seen further, it is by standing on the shoulders of giants'*). At the same time, it acknowledges the incremental and cumulative nature of knowledge production (Schmidt, 1992) and the slow pace of progress that unfolds over knowledge paradigms and generations of theories and researchers.

There are also some cornerstones for slow scholarship. One thing is the approach, to think, rethink, and be both explorative and critical (Ahlqvist & Uotila, 2020). Another thing is to theorize. In fact, the act of *theorizing* lies at the heart of slow scholarship – a process of reflection and inquiry that transcends the immediacy of empirical findings (Hartman & Darab, 2012; Mountz et al., 2015; Bozalek, 2017). Theory in this context is not merely an abstract construct but a condensed practice – a distillation of insights, observations, and reflections that emerge from the crucible of scholarly engagement. It is through the process of theorizing that ideas take root, evolve, and find their place within the broader landscape of knowledge (Meyerhoff & Noterman, 2019). Theorizing is also not just about defining notions to increase precision in how the data is labeled and described. Beyond serving as an index or a vocabulary, it also provides a lens, a way of seeing and interpreting the world, and in doing so, a good theory might also foreground alternative aspects of the world around us. Alternatives that might serve as a call for the collection of additional, complementary, or other types of data.

Central to the idea of *slow scholarship* is also to focus on ways and approaches for letting ideas grow, stabilize, and find their ground (see e.g., Harding, 2008). This is a slow cultivation process that should allow for depth and nuance to emerge over time. It is also a process that serves as a rejection of the relentless pursuit of novelty in favor of a thoughtful approach, and it is a recognition that meaningful insights cannot be rushed or forced but must be nurtured through careful deliberation and inquiry. Slow scholarship serves as a reminder of the intrinsic value of taking time – to ponder, to question, to explore, to critically examine, and to rethink. It is an approach that invites scholars to embrace a more deliberate pace, to resist the pressures of academic hustle culture, and to prioritize depth and rigor in academic work.

Further on, unlike empirical methodologies that are often evaluated based on data, analysis, and generalizability, slow scholarship is measured in terms of theoretical coherence – *does it work in theory*? It is not just about data gathering but also about the construction of concepts, theories, and ideas which are not only able to describe something but also to propose new ways of understanding things and processes, as well as including ways of looking at the world. Ultimately, slow scholarship is not merely a methodology (as an approach for gathering and analyzing empirical data) but more of a philosophy – a way of being in the world – and working as a scholar. It is an approach and attitude that honors the complexity and richness of human inquiry. Slow scholarship, in these terms, is a testament to the enduring power of thoughtful engagement, of ideas that stand the test of time, and of scholarship that transcends the ephemeral currents of the particular moment we live in.

SLOW SCHOLARSHIP, AND ITS RELATION TO "BLUE OCEAN THEORIZING"

"Blue Ocean Theorizing" represents an idea that connects with the principles of slow scholarship, and as such it provides a unique view on academic research. Drawing inspiration from the business strategy concept of "Blue Ocean Strategy" (Kim & Mauborgne, 2011) it proposes making uncontested market space and making competition irrelevant (Kim & Mauborgne, 2014). "Blue Ocean Theorizing" applies a similar approach to scholarly work. Essentially, "Blue Ocean Theorizing" encourages scholars to go beyond the boundaries of existing ideas and limits between subjects, seeking undiscovered areas of knowledge – what can be metaphorically called the "blue ocean," see for instance Grover and Lyytinen (2015). This can either happen through the exploration of a new research direction or through processes of revisiting and rethinking the fundamentals of a particular perspective that has been taken for granted, maybe over decades in a field. This approach is sometimes also referred to as "armchair philosophy" (Williamson, 2019).

For instance, the idea of "human-centered design" has served as one such fundamental perspective in human–computer interaction (HCI) and interaction design research for the last 30 years. The idea has been to explore human–machine interaction not merely from a technological perspective but also from a human perspective where human activities, needs, and motivation are at the center of the design process. This has been a strong perspective in our field, and lots of research has been conducted along this line of thinking. However, with the global challenges we are facing right now, including climate changes, there is a need to rethink this human-centered orientation. Accordingly, the shift to "more-than-human" design represents one such theoretical turn where the human is decentralized, in favor of other values and other species on this planet. The idea is simple, but still radical. We might not be able to solve the current climate crisis only by pushing for alternative solutions for human needs (e.g., "green transition," "fossil free fuel," "green steel," etc.), but we, as humans, also need to take a step back, rethink the position of the human in this larger system, and then formulate other types of applied research projects to move in an alternative, more inclusive, and sustainable direction.

Instead of taking part in competitive efforts to make small improvements within established fields, for instance through research "gap spotting" activities, slow scholars seek to create and explore new paths, challenge beliefs that are commonly accepted, and pioneer innovative frameworks and theories. At its core, this approach to "Blue Ocean Thinking" or "Blue Ocean Theorizing" embodies the spirit of creativity, curiosity, and courage. It serves as an invitation to scholars to commit themselves to research processes characterized by exploration, discovery, and synthesis. Rather than adhering rigidly to established conventions, slow scholars are encouraged to embrace interdisciplinary collaborations and perspectives, drawing insights from diverse fields, and weaving them together into cohesive theoretical frameworks. Moreover, "Blue Ocean Theorizing" emphasizes the importance of vision and foresight in scholarly endeavors (Kim & Mauborgne, 2014). To look at alternative, possible, plausible, or even speculative futures and identifying and speculating on opportunities for transformative change. By adopting a proactive stance toward knowledge creation, scholars practicing "Blue Ocean Theorizing" can position themselves at the forefront of innovation, driving paradigm shifts and shaping the path forward in their respective fields. As such, this slow approach is suitable for the exploration and formulation of new prospects for design research.

With this as a point of departure, a key feature of "Blue Ocean Theorizing" is that it underscores the importance of patience and persistence in scholarly inquiry. While this might sound straightforward, it should be said here that this is not easy – it requires a careful approach, curiosity, time, effort, and resilience. The same is true for the formulation of new theoretical frameworks, paradigms, and prospects. This process is to a large extent about tolerating ambiguity, overcoming obstacles, and continually iterating on one's ideas. Here, it's important to recognize that genuine breakthroughs in our thinking typically only come as the result of cycles of exploration, formulation, critical rethinking, and further refinement.

Ultimately, "Blue Ocean Theorizing" outlines a bold and visionary conception of scholarly inquiry. It encourages researchers to boldly go where none have gone before, to question conventional, established, and taken for granted ways of seeing, to transcend disciplinary silos, and to dream and explore what at first glance seems impossible.

On the other hand, "Blue Ocean Theorizing" is not just about free and brave thinking, curiosity, and explorations. In Sweden, Uppsala University has the motto "To think freely is big, but accurate thinking is bigger" (in Swedish: "Att tänka fritt är stort, men att tänka rätt är större"). In short, theoretically consistent or accurate thinking is hard. As mentioned above, some people might think that blue ocean theorizing is just about so-called armchair philosophy (Williamson, 2019). In other words, that it is to just think freely, openly without restrictions, and speculate about things. However, blue ocean theorizing is as stable as any other form of theoretical inquiry and exploration, but instead of building it bottom up, step-by-step, from data to concepts, it follows a different path, and it should accordingly be measured differently. In the words of Weick, this process is to a large extent about "disciplined imagination" (Weick, 1989) – to think freely on the one hand, but also to critically examine one's alternative concepts, thoughts, and perspectives.

Blue ocean theorizing is also about formulating concepts that describe a particular problem, process, or phenomenon. It is about inventing, formulating, and defining such concepts that, when put to use, make it easier to see, understand, or talk about an (empirical) object of study. The process is typically grounded in an empirical research challenge, topic, or research problem, but it does not seek the answer in empirical data. Instead, it is through a process of trying out and working with words, perspectives, and attempts to formulate definitions that the process moves forward. It is about finding words that speak to what needs to be described, and it is about creating simple frameworks, or so-called "mid-range theories" (Höök & Löwgren, 2012) that explain how these words are interrelated. While empirical data is the test bed for validating a theory, the validation here is more related to the consistency of the framework. In short, *Does it work in theory?*

There are many advantages of blue ocean theorizing as an approach for going slow, not least that it contributes to theoretical development within a field, instead of borrowing theoretical constructs from other fields. But there are also many traps along the way, and many things that blue ocean theorizing seeks to avoid. First of all, blue ocean theorizing seeks to avoid deviation from the object of study by, for instance borrowing concepts and theories from other fields. It seeks to avoid oversimplifications that come from jumping from observations to conclusions, and it seeks to avoid solutionism instead of deepening our understanding of things.

In HCI and interaction design research, we see these traps playing out over and over again. In fact, it is quite common for theories from other disciplines to be borrowed to add theoretical depth to an empirical study (see for instance Berkovich, 2020). In our field, we can for instance see design research papers that present lots of interesting results from the empirical work but then avoid the whole process of theorizing these findings. Instead, research reports are typically about the particularities of the project, lessons learned, prototypes developed, user studies conducted, lessons learned, and ideas for how to move forward. Further on, we can see research where there is a clear path from the empirical work to the design process, and where a new piece of technology is presented as a solution to the identified problem, rather than studies that offer a more in-depth and theoretical understanding of the problem, or more generalized ideas about how that problem unfolds, and how it might be understood, explained, and addressed. And for sure, this is not only established practice in design research, but can also be seen in other areas of research (see for instance Swanson et al., 2017).

As we can see here, blue ocean theorizing does not solve any practical or urgent problems. It is accordingly not a fast or applied approach to research. On the contrary, it is rather slow. The focus is not on fixing anything in practice; it is not about solving anything or providing guidelines, recommendations, implications, or any (design) solutions. Instead, it is about working toward a better or at least an alternative understanding, which of course someone else might find useful. As such, it belongs to the basic sciences, creating new knowledge for the sake of it, not because it is needed for something else.

SOME NOTES ON "SLOW RESEARCH PROBLEMS"

In the processes of academic inquiry and exploration, not all research problems share the same temporal grounds. As I discussed in the previous chapter, some research problems demand rapid answers and fast research processes, while other research problems and approaches unfold at a slower pace, resisting easy solutions and inviting deeper exploration. These are the slow research problems – complex, intertwined, sometimes "wicked," and often enigmatic. In this section, I discuss a set of research problems that typically demand these longer research processes. I start with such research problems and then discuss long-term research challenges, interdependencies, and complex research problems, followed by a discussion on research problems that are big in terms of scale. Finally, I will turn to other aspects that might slow things down, including regulations and other barriers that govern, restrict, or prolong existing research practice.

Wicked Problems – Addressing Complexities

The challenge of solving so-called "wicked problems" – those complicated knots of complexity that resist simple answers – lies at the core of slow research methodologies and approaches to design research (Buchanan, 1992). Wicked problems are difficult to classify and require a complex, comprehensive strategy, in contrast to simple, straightforward, or tame problems that have well-defined dimensions and clear-cut approaches to resolution. Typically, these wicked problems arise in areas where environmental, social, and cultural elements converge, giving researchers a complex set of relationships, dimensions, and unknowns. The wicked problem paradigm is best exemplified by problems like structural inequalities, poverty, and climate change, which demand multidisciplinary cooperation, long-term commitments, and ongoing involvement to fully address and understand.

Long-Term Challenges and Goals – Embracing Temporality

Slow research problems are typically long-term objectives or obstacles that develop over time. These difficulties are inextricably tied to the temporal nature of objects and events, reflecting the social, cultural, and ecological systems' slow evolution, adaptation, and modification. These could include more significant changes in behavior, such as the adoption of more sustainable lifestyles or the alteration of prevailing paradigms in schooling. It takes a patient, forward-thinking approach to work alongside these research problems – one that accepts the fluidity of temporal dynamics and acknowledges the dynamic nature of societal development.

Sometimes it is also the case that this development over time is needed before any descriptions can be formulated or any conclusions can be drawn. To only observe a particular state or a particular moment in time says very little about how something changes, how fast something is spreading, or disappearing, or how well something is adopted or aligned. For sure, some things need to be studied over longer periods, and some things can accordingly only be observed and understood if studied from a temporal perspective – that is, through long-term studies of changes – by definition, over longer periods in time.

Related to this focus, to study things over time is also to set long-term research goals. These goals can then be reached maybe not necessarily in a single study or within a single research project, but over time and over a series of projects geared toward the formulated long-term goal.

Interdependencies and Complexities – Working with Multidisciplinary Perspectives

Interdependencies and complexity abound in the field of slow research are frequently dependent on the advancement of other processes or domains. Emergent, slow, cross-disciplinary tendencies that challenge reductionist methods and demand holistic, integrative modes of inquiry are the result of these connections. Examples of these trends include social justice movements (see for example Fox et al. 2016; Dombrowski et al. 2016), (Öhlund & Wiberg, 2025) sustainability initiatives, and perspectives of more-than-human worlds (Bastian et al., 2016), including responsibilities, responsabilities (Petersmann, 2021) , and care (de La Bellacasa, 2017). A design thinking mentality and a systemic perspective are necessary for researchers working in this area because it acknowledges the interdependence of various phenomena and the necessity of collaborative, transdisciplinary methods to generate knowledge and understanding in these things.

This means that while it might be a good strategy to formulate long-term research goals, there are many other movements to consider in these processes. The research problem might change over time, but there might also be changes that are in one way or another related to your object of study. To keep a keen eye on this means being open to multidisciplinary perspectives and to stay open for how changes and progress made in related fields might also influence your project. While these changes and advancements can be beneficial as they push the field forward, they can at the same time present challenges for the individual researcher who seeks to make a significant research contribution in such complex and ever-changing field of research.

Scale – Addressing Grand Challenges

Another aspect of slow research problems is not only that they tend to be about fundamental questions but it is also typical that the scale is different than more fast-paced applied research processes. In fact, slow research topics are typically of large scale, including technological, environmental, or societal issues that go beyond particular projects, local concerns, specific stakeholders, already established perspectives and approaches, or even fields of research and established research traditions. They might demand seeing things differently (for instance, decentralizing the human, while foregrounding more-than-human perspectives) or thinking about the climate through the lens of "climate justice" (Verlie, 2022).

To truly work toward meaningful change, these "big problems" require coordinated efforts and social actions. The loss of biodiversity, the global health crisis, and the ethical governance of emerging technologies like artificial intelligence (AI) are just a few examples here. For instance, even though AI research is moving at a rapid pace, there are ethical, legal, and regulatory issues that need to be addressed. These issues call for a holistic, thoughtful, and inclusive approach that takes society's long-term effects into account. In short, grand challenges might need to be addressed over a long time, and through long-term, slow and thoughtful processes.

Accordingly, the slow approach might be suitable when the scale of the research problem is high. With small-scale and well-defined research problems, it can be easier to see what approach to take, what methods to apply, and what data to gather. With large-scale and ill-defined problems at the other end of the scale, it might at least in the beginning demand slower processes where the problem is elaborated upon, analyzed, and maybe even divided into subproblems. Just think about the global SDGs formulated by the UN, where many of these problems are not only large scale but also in many different ways interrelated and interdependent.

Addressing such grand research challenges might not only take time, but it can also be rewarding if one manages to do some groundbreaking work. When the research problem is large, and the challenges are grand, the impact can also be fundamental and, if so, then typically long-lasting in terms of research impact.

Barriers, Tensions, and Obstacles – Negotiating Constraints

Design researchers engaging in the slow research approach also encounter a myriad of barriers, tensions, and obstacles – ranging from legal constraints and regulatory frameworks to funding challenges and ethical dilemmas. Legal constraints might involve getting ethical approval for studies that include sensitive personal information and collecting and storing such information over longer periods. They can also involve new restrictions and regulations implemented over time when conducting longitudinal research, and so on. The funding challenges are related to the research funding landscape. Many research projects are typically funded for 3–4 years, are usually applied and oriented toward a certain call, or a specific set of deliverables. If longer studies are needed, it might then demand several approved project applications to ensure that the research can continue beyond one such limitation in time. Furthermore, these barriers may take the form of laws, contracts, or regulations that govern research practices, such as the General Data Protection Regulation (GDPR) in Europe. They may also include broader systemic issues, such as disparities in access to resources or power dynamics within research institutions. To handle and overcome these constraints requires resilience, creativity, building trust, and a commitment to ethical integrity to balance competing interests and uphold the principles of responsible scholarship.

To summarize, I would say that slow research problems demand a patient, thoughtful, and systematic approach – one that considers complexities, temporal dynamics, long-term commitment, and interdisciplinary collaboration. Further on, and if relying on a slow design research approach, it becomes evident that addressing complex, long-term challenges also requires a methodically sound and nuanced approach and ways of using *empirical data* for this long-term purpose. Accordingly, I will, across the following sections, explore a range of suitable empirical methods – from ethnographic field studies and longitudinal studies to participatory design – methods that build on, or are adjusted to, the principles of slow scholarship, offering researchers tools to engage with, for example, user needs, behaviors, and contexts over longer periods of time. Through a discussion of these empirical research methods, I will try to bridge the gap between slow research geared toward theory and slow research defined by its empirical methods (e.g., longitudinal studies). In short, I will discuss how empirical research can be conducted in the context of slow approaches to HCI and interaction design research.

EMPIRICAL METHODS FOR THE SLOW APPROACH

The constant shifting between a strive for rapid insights on the one hand, and contemplative depths of understanding on the other hand, is as I have described in this book an ongoing one in the field of design research. The relation between speed and depth is central to this process, with the slower pace providing a thoughtful counterbalance to the quick pace of contemporary research approaches. The guiding idea of the slow approach to design research is to *prioritize depth above speed*. It requires a thorough and long-term examination of users' actions, paying close attention to every little aspect. Researchers might identify subtleties that could otherwise go overlooked by immersing themselves in the complexities of user experiences through the use of a leisurely approach, as opposed to more rapid-fire approaches that prioritize instant insights. As we saw in the previous chapter, such approaches might include a range of methods, from the quick-and-dirty ethnography (Pink & Morgan, 2013; Twidale et al., 2014) to rapid prototyping (Wilson, 1988).

In this section, I take a specific look at four methods that might be of particular interest when doing empirical work as part of a slow research process. Here, we will take a look at *ethnographic field studies* and *longitudinal studies*, as well as *case studies* and *participatory design*. I also compare these four empirical methods a little bit as a guide for researchers who are thinking about doing empirical work as part of their slow research approach.

Ethnographic Field Studies vs. Longitudinal Studies

The methodologies of ethnographic field studies (see e.g., Randall et al., 2007; Pink & Morgan, 2013; Pelto, 2016; Zilber, 2015) and longitudinal studies (see e.g.,. White & Arzi, 2005) not only serve as empirical tools but also embody philosophical principles that align with the concept of slow scholarship. Let us look deeper into the philosophical underpinnings of these methodologies and their resonance with the core idea of slow scholarship.

At its core, ethnographic field studies focus on immersion, engagement, and deep contextual understanding – the very essence of slow scholarship. Researchers, akin to anthropologists, seek to go into the natural habitats of users, not merely as observers but as participants in the fabric and emergence of everyday life. This immersive approach, sometimes described as "going native," is grounded in the philosophy of phenomenology (Qutoshi, 2018), which emphasizes the subjective experience and the importance of social and physical context in shaping human behavior. From a philosophical standpoint, ethnographic field studies align with the principles of reflexivity and situated knowledge (Burkitt, 1997). Researchers acknowledge their role as participants and co-creators of knowledge, recognizing that their presence inevitably influences the dynamics of the research context. By engaging in participatory and reflexive inquiry, researchers applying this method need to balance the complexities of subjectivity and bias and strive to uncover perspectives, values, activities, and rituals that resonate with the lived experiences of users. Moreover, ethnographic field studies embrace the hermeneutic tradition (Ormiston & Schrift, 1990; Smith 1991), viewing human behavior as imbued with meaning and interpretation. Researchers who apply this method adopt an interpretive stance, seeking to understand the underlying meanings and symbolic interactions that shape user experiences. This interpretive approach fosters empathy and compassion, allowing researchers doing ethnographic work to connect closely with users on a deeper level and generate insights that transcend surface observations.

In contrast, longitudinal studies (White & Arzi, 2005) offer a temporal dimension to the exploration of human behavior, echoing the rhythms of slow scholarship. By tracking users' behaviors and preferences over an extended period, researchers can study continuity and change, akin to the philosophical concept of things that are slowly becoming, unfolding, transforming, vanishing, growing, or escalating. Longitudinal studies embrace the temporality of human experience, recognizing that behaviors are not static but evolve over time, and in response to myriad influences. From a philosophical perspective, longitudinal studies resonate with the principles of process ontology (Di Poppa, 2010) and relationality (see e.g., Houseman, 2006; Bottero, 2009; Tynan 2021). Researchers doing longitudinal work view human behavior as dynamic and relational, shaped by interactions with social, cultural, physical, and environmental contexts. This relational view emphasizes the interconnectedness of phenomena (Linklater, 2010; Acquaviva, 2000) and underscores the importance of context in shaping human actions and experiences. Furthermore, longitudinal studies embrace the principle of epistemological humility, acknowledging the limitations of linear causality and deterministic explanations. Researchers conducting longitudinal studies adopt a nonreductive stance (Hollway, 2010), recognizing the inherent complexity, unpredictability, and dynamics of human behavior. This humility invites researchers to embrace uncertainty and ambiguity, and it typically results in a research process that seeks to move beyond simplistic narratives and instead embraces the richness of histories, memories, stories, and lived experiences.

If we now relate these empirical methods to the philosophical ideas underpinning slow scholarship, we notice a shared ground rooted in depth, engagement, and reflexivity. Ethnographic field studies and longitudinal studies offer complementary perspectives on understanding human behavior and how it changes and plays out over time, each relying on philosophical principles that align well with the values and fundamentals of slow scholarship.

When we as researchers conduct these long-term studies where we "go native" and become part of the context we study, we are not mere observers but active participants in the process of knowledge creation. By embracing the philosophical underpinnings of ethnographic field studies and longitudinal studies, we also cultivate a deeper appreciation for the complexities of human experience and the importance of context in shaping our understanding of the world. In short, we blend in over time, and through this process of blending in, we both affect and are affected by the context we study – over time.

Case Studies vs. Participatory Design

If we now move to the two methods of case studies and participatory design methods, we can notice how these two strands of empirical methods emerge as not only empirical tools and approaches, but also how these two embody philosophical ideals that resonate with the core idea of slow scholarship.

If we start with *case studies*, we can see how this method offers an approach for exploring the particularities and intricacies of user experiences within a particular social context – a reflection of the narrative tradition within slow scholarship. Researchers become not only observers but also storytellers about user interactions, needs, and preferences. This narrative approach embodies the principle of narrative inquiry (Savin-Baden & Niekerk, 2007) – an approach that recognizes the power of stories to shed light on human experience and reveal deeper insights about social interplays. Philosophically, case studies as a method align with the hermeneutic tradition, viewing human behavior as imbued with meaning and interpretation. Researchers who are engaged with case studies typically adopt an interpretive stance, seeking to uncover the underlying meanings and symbolic interactions that shape user experiences, activities, rituals, and structures within specific social contexts. This interpretive approach enables researchers to engage in dialogue with the narratives of users, exploring the nuances of their activities and lived experiences. Further on, it seeks to describe, interpret, and uncover insights that transcend surface observations.

Participatory design (PD), on the other hand, embraces the principles of active engagement, collaboration, co-creation, and empowerment (see e.g., Holmlid, 2009; Nordin et al., 2023) – these principles so fundamental to the PD approach form the hallmarks of the relational ontology that serves as a core foundation of slow scholarship. In this method, the participants are not passive subjects but active agents in the design process empowered to inspire each other and co-create solutions alongside researchers and designers. This collaborative dimension of participatory design methodology embodies the principle of relationality, which recognizes the interconnectedness of individuals and the importance of shared agency (Ertner et al., 2010) in shaping collective outcomes. Philosophically, the method of participatory design aligns with the principles of epistemological humility and democratic engagement (Thinyane et al., 2018). Researchers engaging in research projects through the method of participatory design adopt a stance of humility, recognizing the expertise and lived experiences of the participants as valuable sources of knowledge. By inviting people to become active collaborators, participatory design democratizes the research process, ensuring that solutions resonate with people's real-world activities, needs, values, and priorities. In the context of design research, it is an example of participatory design aimed at supporting indigenous people and empowering them through design (see Nordin et al., 2023).

If we now relate these methodologies to the philosophical ideas underpinning slow scholarship, we should notice a shared focus on depth, long-term engagement, and empowerment. Case studies and participatory design offer complementary lenses through which to examine user experiences, each in line with philosophical principles and standpoints that align with the values of slow scholarship. In these approaches, we as researchers are not mere observers but active participants in the process of knowledge creation. This also illustrates that slow approaches to research do not only need to be theoretical work. On the contrary, there are several empirical approaches to explore while keeping a slow and long-term perspective. By embracing the philosophical underpinnings of case studies and participatory design, design researchers can cultivate a deeper appreciation for the complexities of human experience and the importance of collaboration and co-creation in shaping meaningful designs, long-term solutions, and groundbreaking impact.

Advantages and Considerations

The slow approach to empirical design research brings forth a multitude of benefits, each contributing to a richer understanding of user needs and behaviors. The slow approach offers not only a greater understanding of social context and insights into how digital technologies might be embedded in social settings, but it also foregrounds interpretations and interdependencies. As such, the slow approach enables HCI and interaction design researchers to uncover social, cultural, and environmental factors that influence and shapes user experiences.

However, the slow approach to empirical design research also comes with a set of challenges. By definition, the slow approach demands time, resources, and long-term determination and commitment, and it may not always be suitable for every research endeavor. To have a good understanding of when to select the slow approach to a particular research challenge requires careful consideration of the research aim, context, and goals, as well as the scale and urgency of the research problem. In contexts where complexity demands depth and contextual understanding, the slow approach offers a lens into the particularities and intricacies of human activities and experiences. In dynamic environments where user needs and preferences are subject to change, the slow approach's long-term, iterative, and adaptive nature enables researchers to refine their strategies in response to emerging insights and evolving contexts – over time.

CONCLUSIONS FROM GOING SLOW, AND MAKING LONG-LASTING IMPACT

The slow approach to design research serves as a reminder of the benefits of patience, depth, and calmness in a world that frequently favors speed, efficiency, and deliverables. By adopting a slow approach, researchers can find insights that strongly resonate with users' actual experiences by adopting approaches that stress comprehension and emergence over urgency and immediacy. This opens the door to design explorations that are not only successful but also explorations that are challenging and explorations that potentially might have a long-lasting impact.

In the dynamic field of design research, the slow approach accordingly includes several empirical methods – ranging from case studies and participatory projects to longitudinal investigations and ethnographic field research – approaches that enable a connection between slow empirical work and processes of theorizing. To slowly gather data and use this collected data to fuel and develop alternative ways of seeing the world is an important part of the slow approach.

But beyond the empirical methodologies that I have tried to cover in this chapter lies a deeper approach, a philosophy that underpins the slow approach to design research. It is, as I have described here, a philosophy rooted in patience, curiosity, engagement, and humility. A philosophy that acknowledges the complexities of human experience, the limitations of our understanding, and an approach that enables us to stop and reflect, rethink and form a foundational understanding of things. The slow approach is also a recognition that true insight emerges not from haste, but from the willingness to linger in uncertainty, to embrace ambiguity, and to allow the seeds of ideas, speculations, and understanding to take root, grow, and flourish – over time. Although this might sound a bit poetical, it is exactly that poetic approach that is needed when questioning and rethinking what we think we know, and when seeking alternative ways of seeing things.

As I have pointed out in this chapter, the slow approach to design research is not without its challenges. It demands time, resources, long-term goals, and unwavering commitment. It requires a balance between rigor and relevance, between depth and breadth, between immediacy and longevity. Yet, it offers a promise – a promise of depth, of richness, of reflection, of different perspectives, of abstractions, of precision in wording, and a promise of enduring impact. It's a promise that resonates with the very essence of design research – to illuminate the human experience, to inspire innovation, to rethink current ways of doing things, and ultimately to shape an alternative, and better, world. In short, the slow approach offers an alternative way of seeing – through design, and over time.

We now have an understanding of the fast and slow approaches, so in the next chapter we take this as a point of departure to look at how the paces of HCI and interaction design research unfold – over time.

CHAPTER 4

Paces of Design Research – Methods and Perspectives

INTRODUCTION

Central to design research is the idea of moving forward. Exploration, change, development, and progress are fundamental cornerstones of design research, and by definition, such transformational progress unfolds over time. Through these processes, we explore topics, settings, and problems with the intent of finding new solutions, imagining alternatives, or establishing new ways of seeing. Still, while we're steadily moving forward, the pace of our research process shifts – related to the activities we undertake, related to the methods and approaches we select, and related to *when* we intend to make a research contribution. As I stated at the beginning of this book, this is a relational concern in terms of how we arrange the research process in relation to the contribution we are aiming for, and as a result, the pace of the design research process changes accordingly.

Sometimes the pace is *slow*. We might need to dig deep into a design brief, we might need to spend time to study, understand and rethink a particular setting, or we might need to spend time and work hard to address complex and sometimes wicked problems. Such activities take time. We might also need to rethink, reevaluate, and reimagine, or we might need to iterate, revisit, and reorient. Again, these processes take time.

DOI: 10.1201/9781003343745-4

In other cases, the pace is *rapid*. We elaborate, we sketch on a piece of paper, we brainstorm, we do rapid and low-fi prototyping, and we ideate. This is of equal importance as the slow and steady movements in any design research project. We need to let our minds be free, to just brainstorm and generate ideas, and we need techniques to collect all these ideas. Sometimes we use post-its, and sometimes we just sketch, draw on whiteboards, or simply chat about it.

In fact, the idea of moving forward and being future-oriented in this approach has always been central to design research. As formulated by Herbert Simon (2019) in his book *The Sciences of the Artificial* this type of research is about "moving from an existing situation to a preferred future situation." In this statement Simon pinpoints a "now" as the existing situation, a future situation (a future now), and a research process that should constitute the transformation of the existing into one such preferred future situation.

But how far away is this preferred future situation? How long into the future should we aim with our design research projects? And how likely is it that we can arrive at this future state through our research efforts? For sure, if we only imagine tiny changes, then maybe the time frame can be short, but if we imagine a radically alternative future, then even a longer time frame might not be enough. In addition to this, we should acknowledge that it is not solely a particular design project that changes our reality. In fact, lots of different factors continuously change our everyday, and it is the myriads of small efforts taken that collectively form trends and streams of change in society. So how should we pick a suitable time frame, and an appropriate pace for our design research projects to ensure that we are capable of reaching a preferred future situation within time, and is that what it means and takes to be timely?

In this book I address this spectrum of paces in design research – from the rapid activities we undertake to the longer periods of time where we seek alternative ways of seeing things. Still, while there are many methods and approaches developed for doing design research, little is said about the rhythms and paces that go into these processes, or how to make timely contributions.

In fact, it is hard to do timely design research. As illustrated in Figure 4.1 the research problem is defined at a particular moment on time (t1), but the contribution can only be made ones the project is completed, i.e., at a later stage (t2). The paradox here is that it is easier to be timely if doing fairly short (and accordingly quite limited) design projects, but then the

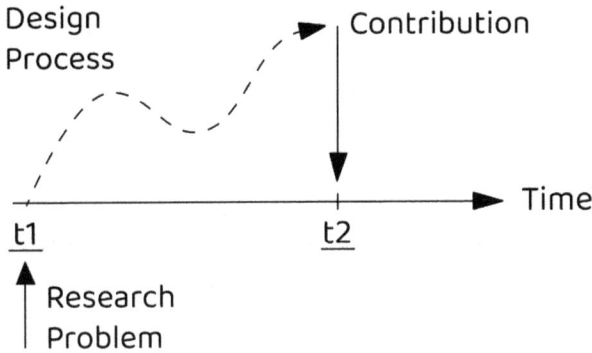

FIGURE 4.1   Design research process over time – from initial research problem to contribution.

problem is that the contribution from such limited projects might, in fact, be equally limited.

On the other hand, a very long design research process might deviate too much from the initial problem, or the field might have taken turns that make the design project irrelevant. The problem might have changed, or the need for solutions might not be there anymore. This is something I will come back to at the end of this chapter. Indeed, this has to do with how to manage to do timely design research, how to make contributions to a constantly evolving body of existing research, and how to contribute to an ever-changing practice.

In this chapter, I reflect on these different paces and approaches to design research, and I discuss how the approaches of "going fast" vs. "going slow" come together in design research.

Here I also try to break away from the stereotypical understanding of fast vs. slow, where "fast" is thought of as a highly collaborative process, with relevance as a key dimension vs. descriptions of the slow approach as a highly individual and socially isolated effort, with rigor as a key dimension (a seeming dichotomy typically being articulated when discussing applied research (as a social, collaborative, and action-driven activity) in relation to basic research (which is typically described as being about the individual who seeks to dig deep, analyze, and theorize). Beyond these two extremes, I suggest that research is a complex activity that involves both collective and individual efforts – no matter if it is more of an applied or basic research project. Research typically involves an interplay and collaboration between the many individuals who are coming together in a research project. In practice, this social interplay can take many forms, such as research groups, joint research efforts, work packages in larger projects, and co-authoring grant proposals and research papers.

However, recruiting participants, obtaining ethical approval, and navigating a constantly changing research field all require flexibility and planning in the research process. This can lead to tension between the need for fast solutions and the value of slow, thorough research – and it can lead to tensions in research groups. Given a need to act (fast!) – the things which demand careful and systematic processes, and as such demand more time, can be frustrating to carry out, if still needed in the research process. Accordingly, I discuss this as one dimension of how the fast and slow comes together in practice.

Fast approaches to research outputs, such as the publication of short papers, the presentation of late-breaking results, arrangements of small-scale workshops, or ideation with low-fi prototyping, can be useful for trying out design alternatives and iterating on ideas. As Beck and Stolterman-Bergqvist notes, "there are no clear problems, no clear methods, and no clear solutions in designing" (Beck & Stolterman-Bergqvist, 2018). This highlights the importance of combining an interest in moving fast with some slow approaches that allow for deeper reflection and consideration of complex issues. When dealing with these uncertainties, there is a need to quickly explore things, and then also provide the time necessary for making good decisions on how to proceed.

The COVID-19 pandemic between the years 2020 and 2022 can serve as an example of a situation where fast solutions were urgently needed – including the development of a vaccine, treatment methods, and solutions related to new ways of working online, or staying in connection with friends and family during lockdown, and at a time where everything had to be moved online (Wiberg, 2020).

On the other hand, other issues such as those addressed in ACM Interactions' theme issues on gender, climate challenges (Wiberg et al., 2022), race (Harrington et al., 2021), disabilities (Rosner et al., 2021), and inclusive design require more long-term, thoughtful, and critical approaches. For these large-scale issues, there might not be any quick fixes or fast solutions. In fact, it might be the case that there cannot be any such solutions to these grand challenges. There is no way of just adding a solution to the problem. Instead, the whole system, structure, or society might need to be reimagined and reconfigured to address these grand challenges. These problems associated with solutionism have been discussed in a ToCHI paper by Alex Taylor and Daniela Roser (Cunningham et al., 2023) where they highlight the limitations of fast, one-size-fits-all solutions, and as such this article serves as a call for more long-term, fundamental, and groundbreaking research ambitions, goals, and prospects.

In this chapter I also cover how difficulties in the research process can lead to slow or delayed research, which is different from the intentionally slow approach discussed here (i.e., deviation from the "going fast" approach, and its results in "going slow"). Here I suggest that it is important to carefully consider the appropriate pace for each research project, and to balance the need for fast solutions with the (maybe long-term, or more groundbreaking) value of thorough, and thoughtful design research processes.

THE PHASES AND PACES OF DESIGN RESEARCH

As an introduction to this topic, let us take a look at the well-known "Double Diamond" model that has been extensively applied (see for instance Kochanowska & Gagliardi, 2022; Saad et al. 2020) although this model has also been heavily criticized as a structure for, and representation of, the design process. The Double Diamond model is a visual representation of the design process, first proposed by the British Design Council in 2005. It illustrates a design methodology that is split into four main stages – Discover, Define, Develop, and Deliver – across two "diamonds" (see Figure 4.2). The first diamond represents the problem space, and the second represents the solution space.

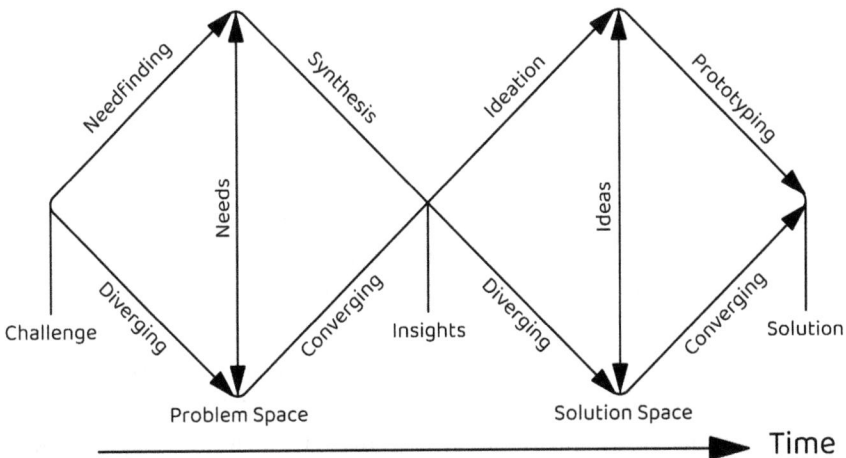

FIGURE 4.2 The "double diamond" model – four phases of design processes (with assumed equal duration/pace), typical/ideal rhythm of design research (across four phases) if viewing this as a linear model over time.

Typically, the first two phases of the Double Diamond model pertain to research about the problem space, whereas the following two phases are about the solution space. As a design researcher, you are expected to go through these four phases – from the initial "design challenge" to the final design. Still, nothing is said about the length or pace of each phase in this model. For instance, to understand the problem space might take a long time and might involve extensive data collection and analysis. There might be several layers to understand, from the material needs to more complex matters that need to be understood (e.g., social dynamics, structures, culture, politics, etc.). This might take lots of time. Also, if we consider the second phase here of ideation and prototyping, this might not be a fast, linear, and smooth process. In fact, to this date, there is no methodological support for moving from an understanding of an existing situation to imagining a future preferred situation. Or in other words, you cannot derive the "ought" from the "is" – there is no support for moving from descriptions to normative statements on alternative future states. In this transition from "what is" to "what could be," or "what ought to be," we need to rely on a set of techniques to generate new ideas. We sketch, we assemble mood boards, we brainstorm, and we run workshops. Ideation might go very fast, but it might also take a very long time – it depends.

ON DIFFERENT PACES ACROSS THESE FOUR PHASES

The "Double Diamond" model gives us the impression of a clear, clean, and well-structured design process across four phases, where each phase is of similar length in time. However, in reality, it might be very different. For instance, data collection might be fast, but the analysis takes a long time, followed by a rather short period of ideation and prototyping (Figure 4.3, A). Or the data collection takes a long time (e.g., a longer ethnographic field study) (Figure 4.3, B). It can also be the case that the problem space is easy to understand and, accordingly, can be done rapidly, but to iterate and build a suitable prototype that really solves the problem identified might be complex. Due to this complexity, the final phase might take a long time (Figure 4.3, C).

In this model, it is also assumed that the design researcher always reaches the fourth "deliver phase." That might not always be the case. Sometimes one of the previous phases fails – for instance, due to lack of data or access to the most relevant data. Another reason for why it sometimes fails is simply because the research is no longer timely, and accordingly no longer needed or relevant. Circumstances might have changed;

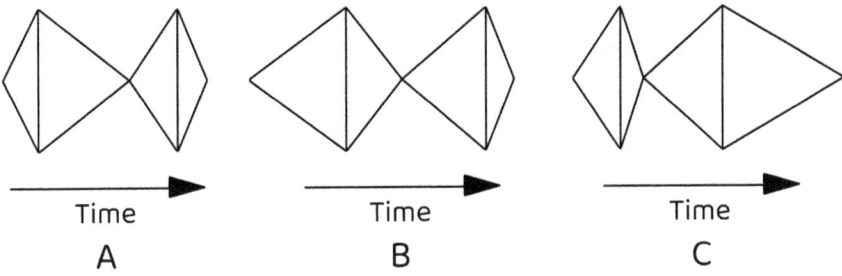

FIGURE 4.3   Moving with differences in pace across the four phases of design.

there might already be alternative solutions available, or there might already be related research published that addresses the problem space. To be timely is accordingly a relational dimension of the design research process.

ON MOVING THROUGH DESIGN – AND AIMING FOR A TIMELY CONTRIBUTION

Now, let us examine this idea of making timely research contributions a little bit closer. For instance, what if there are new emerging areas? Or what if an area is gradually disappearing or being replaced by another research area? In Figure 4.4 we can see a research area that stretches from t0 to t1, and we can see how a slightly different area is emerging during

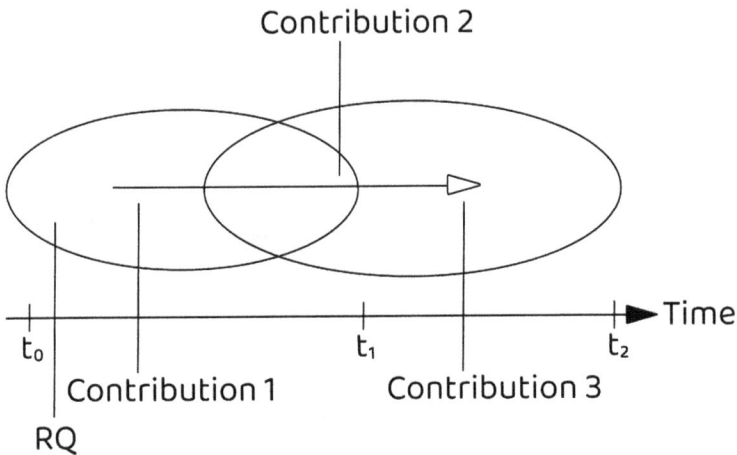

FIGURE 4.4   Emerging and evolving research areas – over time, and the challenges associated with making timely contributions in this ever-changing research landscape.

this first phase, and how this second area ends at t2. Now, imagine that you start up a new design research project where you formulate the research question ("RQ" in Figure 4.4) in relation to the existing body of research and in relation to this growing body of contemporary research. You can of course go for a short project and seek to make a quick contribution (contribution 1). This might ensure that your project is timely, but it might not be extensive enough to constitute a fundamental contribution. The time is also short as there is a new perspective or related area that has started to gain some interest in the research community. Another option then is to aim for a longer project. Here, the risk is that the contribution needs to be made before it is no longer relevant from the viewpoint of how the initial project was defined in relation to that initial research area (contribution 2). An additional risk is that the second area has now been established, so it might be more relevant to contribute to this new area of research, but then the risk is that the project is not fully situated in or aligned with this new area. Finally, you can also go for more long-term projects that stretches across existing and future areas, and you might then aim for a contribution at a later stage (contribution 3). This approach also has its challenges as the research problem, the approach, and the theoretical lens might be outdated when it is time to present how your research adds something new to the established body of existing research.

So how should we deal with these challenges? And what strategies can we apply to ensure that we do timely and relevant research? Here I would like to suggest the two most common approaches taken, that is, to either 1) *be quick*, to move rapidly and to adjust the research agenda in relation to these changes in the research landscape – this is the fast approach to design research, or 2) *to steadily and systematically move forward* – that is the slow approach to design research. Across the following two sections I describe these two approaches, and I relate these two approaches to the notions of applied vs. basic research and what the two approaches can offer for design research in general.

THE PACE OF "GOING FAST" – ON "GAP FINDING" AND QUICK CONTRIBUTIONS

Design research is about generating new knowledge on ways of making change happen – through design.[1] As such, design research is an activity of making change happens – in practice, and through design, while at the same time generating new knowledge. Of course, existing practice is about the "here and now," but in order for design research to not only be about design but also be about research, it needs to be related to how the research field is evolving, over time. To make a research contribution through design thus becomes this dual act of both contributing to practice while at the same time aligning this with a timely, and hopefully growing strand of published research.

It is quite easy to see how a strictly applied approach to design research both can be rewarding and dangerous at the same time. If going for an "applied approach," then of course research becomes an act of "applying" what we already know to a new problem, and the focus will then be on the outcome of these processes of "applying" what we know to a new context. This can indeed be rewarding as it allows for a quick study of the current situation, and quick use of existing frameworks, models, methods, technologies, and techniques for addressing and maybe even solving the problem at hand. It is rewarding in that no new approaches or frameworks need to be developed. Instead, it is possible to maintain a focus on the outcomes of this process in terms of results, and it is likely that the process can be fairly short. However, the *"gap finding and applying what we have"* approach to design research is also dangerous as it will only look for existing "gaps" in what we already have, rather than seeking to rethink current practice and existing strands of research, and challenge and expand our research in new directions. It is also focused on these "gaps" where work is lacking, rather than going for reinterpretations, new approaches to, or alternative theoretical and critical reexaminations of larger portions of our field. In short, this approach allows for "filling the map" research while simultaneously turning away from more in-depth or critical studies that might lead to more groundbreaking research.

Still, there are good reasons why many researchers favor the applied approach. We are in a field of research where lots of things change at a rapid pace. There are constantly new materials and technologies being developed and made available.[2] At the same time, there are constantly changing needs and societal challenges. The design researcher needs to identify a sweet spot here between a timely topic and a growing interest in an ongoing development.

As an example, there is at the current moment a growing interest in what has recently been referred to as the 3rd wave of artificial intelligence (AI), where this third wave can be characterized by efforts made to apply AI to societal contexts (Xu, 2019). At the same time, when this third wave was at its early stages, there were still few documented examples of applying AI to societal contexts, so in the beginning of this wave, there was an opportunity to conduct design research that explored applied AI in real-world social contexts. However, with the established gap spotting approach to research this wide-open gap (of "few documented examples") has been rapidly filled with publications where researchers are reporting experiences, and lessons learned, from applying AI technologies in various societal contexts.

For design research this is again both an opportunity and a threat. These trends or "waves" come and go. There is always a new wave of developments, and there are constantly new technologies being developed. However, to make a contribution, the design researcher needs to act fast. If the research problem is formulated as a gap, or a lack of studies, then the research is only of interest as long as this lack or gap exists. That is, until there is some work published. This suggests that the research process needs to be fast, or that the gap is either irrelevant to others, or not known by others.

From this perspective "gap finding" and research aimed at addressing "gaps" is a dangerous approach. As suggested by Whittaker et al. (2000) "we need to stop pushing the envelope and start addressing *it*," or in other words, we cannot just focus on the latest technology, or "the next big thing." We also need to look beyond just developing new technologies and applying those to new social contexts. Instead, we need to start "addressing it," that is to find ways of articulating and approaching the more fundamental dimensions and questions of the things we explore in our design projects. In short, to not just look for the gaps, but also rethink, iterate, and reexamine the things we think we know.

A focus on "the latest" also, and almost by definition, demands a fast approach (or at least a faster approach than any competing research initiative), and one such approach does not intuitively promote open research collaborations (due to the risk of revealing a gap to others before it is addressed, published about, and accordingly "filled" or at least addressed). Still, the fast approach is quite popular. It is easy to spot a growing interest in a new technology, and it is relatively easy to contribute initial studies to new emerging application domains, or to contribute with examples of how a new technology can be used or applied. Further on, is quite easy to see how contributions can be made by focusing on new emerging technologies, and to do research on interesting application domains. For instance, as we now have a focus on emerging technologies like 5G and 6G mobile networks, Internet of Things (IoT), and artificial intelligence (AI) we can probably make some contributions on for instance "the use of AI in healthcare," or "the use of 5G or IoT technologies in logistics." In this way, we can contribute by *applying* a new technology to an existing application domain, profession or social context. We can also contribute by *combining* new technologies, e.g., by using 5G and IoT in the logistics context. We can also make fairly rapid contributions here by tackling new technologies with somewhat eternal questions and perspectives, i.e., to contribute by *questioning.* For instance, by running a research project on "AI and ethics," or "IoT, surveillance, and privacy."

To summarize, we can view the fast approach to design research as a matter of trying to make a contribution fast, to contribute to an existing situation before any competing initiative fills that particular research gap. We can do this by applying something new, say a new piece of technology, to a well-known social context (for instance "AI in healthcare"), or we can tackle a new technology with traditional and well-established perspectives (for instance "AI and ethics").

To move fast, we also need to make sure that the "Double Diamond" design process is short enough to be conducted within a limited time frame. This can be achieved by working on a well-defined problem, with well-known methods and technologies. These research projects are typically about making minor contributions, taking smaller risks, and staying focused on applied research – and accordingly such contributions are typically related to applying or testing out something well-known in a fairly well-known problem space.

To summarize, fast is typically about:

- Needs and pressing problems
- Application areas
- New technologies
- "the applied"

What is needed? – Developed methods and frameworks that can be applied

THE PACE OF "GOING SLOW" – ON MAKING LONG-TERM CONTRIBUTIONS

Being fast is not the only approach to the making of research contributions. Being fast can imply more direct impact, but the topic might also be interesting for a quite limited time. Different research topics have different lifespans, and for the fast approach the typical topic is also quite limited in this respect – the researcher needs to move fast, for multiple reasons.

There are of course alternatives here. If the focus is on "AI and ethics" then the approach taken to this topic can in itself determine if it will be a fast or slow project. If the focus is on analyzing the ethical aspects of a particular AI implementation, then the process can be relatively fast – as this is a process of using ethics as a tool or perspective to analyze a particular technology and how it is used (i.e., an applied research process). But we can also shift this around. In the beginning, AI research was to a large extent devoted to building "intelligent machines," and in doing so a more long-term goal was to learn more about human intelligence, reasoning, judgment, the brain and the human mind, and it was about understanding how cognitive processes work, through the making of these "intelligent" machines. If we think about "AI and ethics" from this perspective an alternative, and a more long-term research agenda could be to use a focus on AI as a vehicle for understanding ethics better, i.e., if AI is a manifestation of an "ethical machine," then how does it need to be further developed, and what new ethics do we need to imagine, articulate and design along, when intelligence is no longer restricted to individuals and their communicative capabilities, but extended to large-scale high-speed clusters of interconnected computers? In short, how do we need to reimagine ethics in the age of AI? This is a larger, and more long-term question that also touches upon some basic research questions (e.g., "what is ethics? To beging with"). This alternative focus demands more time, and accordingly, longer research processes.

There are also almost eternal research topics, and I will return to this toward the end of this book. For instance, "sustainability" is one such topic. In fact, the topic is almost a truism – how can we not consider sustainability if thinking about the long-term aspects and effects of the design research we conduct?

In addition to these examples there are also other topics that are long-lasting due to their recurring nature. For instance, how design not only enables a future situation, and empowers some people in this transition, but also how it also changes the existing situation. These transitional dimensions of design will always be a topic for any design research project as this will surface in the research process over and over again.

Another aspect of "slow" is that research can never happen in isolation. Knowledge production is not only about facts and results, but it is also a social process, from how we conduct research in research groups and research collaborations, to how a research community work collectively with reviews, questions and ways of challenging the existing knowledge base, the current research front, or what is at the time being considered as interesting and relevant. In this context we typically describe this work as community service, and we think about this as "academic citizenship."

Over time, any research field changes, and this is a slow process. Further, and in adding to this, we can think about how we add contributions to an existing body of design research. We do this individually, or together with fellow researchers, but its typically through projects, and through written and published papers. The projects and the writing of these papers take time. In practice, this mean that the process – from an initial idea, via a design research project, and via writing, reviewing, revisions, and publishing – can take several years. From the perspective of slow this is good – it is important that we focus on research challenges that are relevant not just now, but hopefully for many years to come, and it is important that we have this friction in the system to ensure that we are careful about how we challenge, develop or even changes what we consider to be an established body of knowledge.

This can sometimes stand in contrast to short research projects, calls for special issues, or other pressing deadlines. At the same time, this is why it is important to contrast the fast and applied approach to research with a perspective that illustrate why slow approaches are needed if also aiming for research with long-lasting impact.

To summarize, we can view *the slow approach* to design research as a matter of striving for more long-lasting results. It is about contributing to alternative ways of seeing (theorizing) or approaching (methods) an existing situation. In slowly moving forward we might need to go through multiple "Double Diamond" processes, as to gain experience and collect data of how things develop – over time. In the slow approach to design research we work on ill-defined, unexplored, or wicked problems, with new, or fairly unexplored methods and approaches. These slow- or long-term projects are typically about seeking to make larger contributions, taking higher risks, and to stay focused on basic research – and accordingly it is about seeking ways to contribute to our fundamental understanding of things.

Slow is typically about:

- Gaining understanding rather than finding solutions

- Shifts in thinking rather than "more of the same"

- The basic (fundamentals)

How can we do this, in a timely way?

- Understand growing areas and perspectives
- Understand and be sensitive to new trends and trajectories, i.e., signs of movements within the field

FROM CONTRIBUTIONS TO MAKING PROGRESS THROUGH DESIGN RESEARCH

We can probably not do with either the slow or the fast approach to design research. Beyond any such separation or hard choice, we need to find ways of working across both timely topics, and more everlasting, or at least recurring questions. In other words – we need to find ways of linking the now to the everlasting. For just about any scientific area this is called *"progress"* – *we move forward, we look back, we revisit and we challenge, and then we move forward again.* So, what is progress then?

Progress, here understood as the overarching pace of a scientific field, is another fundamental cornerstone of design research. On an overarching level we move forward collectively by making single contributions to a shared body of design research. This body of research consist of studies, examples, and development of conceptual frameworks, and it consists of methods and experiences from applying these methods to design projects. Further, it consists of the actual design work we do, and the documentation of design projects, including sketches, scenarios, and prototypes. Further on, it consist of explorations of alternative futures, and ways of articulating these attempts to speculate about what the future might hold for us. In short, all these efforts move our field forward, and it is through the making of all these different efforts, through all these contributions, that we are moving design research forward. Of course, all these different activities take more or less time, and as we bring these efforts together, we can notice how even rapid contributions to a fast-growing strand of research is simultaneously part of a slower movement, or even part of a larger slow-moving shift in a field. In this way the aspects of fast and slow research are always intertwined, and this is why the paces and rhythms of design research is a complicated matter to deal with – especially if trying not only to do design research that is timely but also to do design research that matters – in the long run.

Ultimately, we pick timely research topic where we can conduct a "design size" project and 1) contribute with some timely results, and 2) relate the particular findings and results to a larger movement in society, and within our field of research. This demands an understanding of movements in our surrounding society – to have a sensibility for pressing and urgent matters (i.e., contemporary research problems), and it demands a capability to see and understand the slower movements in the existing field of design research (i.e., trends, and typically slow, movements, changes and transitions in methods, areas, and conceptual/theoretical framings and frameworks).

As a field interested in *change*, which is of course a central and fundamental component in design, we are also moving forward by chasing a number of running targets. The research object in design research is very seldom stable – on the contrary, it is typically unstable, and dynamic, and it also changes over time. This is not only due to changing pre-conditions for design, but it is also due to the development of new technologies, and the application of new technologies to new application domains, changes in expectations, changes in peoples understanding and use of new technologies, and so on.

In adjusting our design research agenda to these changes, we have some things and some different approaches to consider. For instance, if we are constantly chasing running targets, we need to 1) understand what these different existing and emerging targets are, 2) we need methods to identify emerging targets, and 3) we need to know how to follow these targets as they change – over time. In short, this approach is an approach that seeks to understand contemporary and emergent issues, and it seeks to create new knowledge about these emergent trends, and to identify "the next big thing." As an approach, it is well-suited to work with timely topics. On the other hand, that is also its main disadvantage – what is timely at one particular time, can easily be outdated as soon as "the next big thing" is identified. In our field of HCI and interaction design research, we see this over and over again. One year it is a huge focus on "IoT – the Internet of Things." The next year the focus may have shifted, maybe to "5G mobile networking," and all of a sudden, the focus changes again – maybe to "applied AI" or any other overarching technological trend in our society.

Another approach to "progress" is to combine a keen eye on how practice changes with an eye on what is relatively stable over time. I like to think about this approach as a *theory-driven approach* to design research (Stolterman & Wiberg, 2010). In being theory-oriented I do not suggest that we should rapidly develop new theories. On the contrary, being theory-oriented is about moving slowly but steadily forward, and to revisit the basic concepts and notions we have, and its definitions and meaning in relation to changes in practice. One such approach also creates stability in an otherwise fast-moving field. Here we can for instance revisit and reexamine such basic and fundamental questions as for instance, "what is design?", "who are we designing for?", or "what is interaction?" By asking such fundamental questions, and by revisiting and reexamining these questions over time we make sure that we refine our definitions, we refine our (theoretical) understanding (of practice), and we make sure that although we are moving slowly with this development, and across multiple empirical cases and inputs from practice, our theories are aligned with how practice develops.

Hopefully this serves as an illustration of how the different temporalities operate simultaneously, and as an illustration of the importance of aligning efforts on long-term theoretical development with a more flexible sensibility for the dynamics and rapid changes in practice. To be theory-oriented is accordingly a concern for how the area of design research develops (over time), and in combination with being practice-oriented it is simultaneously a concern for being relevant to the ever-changing dynamics of practice.

My advocacy here for a theory-oriented approach to design research is accordingly not about expanding our field in terms of moving in new directions, but about deepening our field. As formulated by Marcel Proust, "The real voyage of discovery consists not in seeking new landscapes but in having new eyes." Following from this, It is not about generating lots of new concepts for new things, but about revisiting our core as we move forward. In choosing this approach I see it as an alternative path for design research that provides an alternative to the contemporary "gap spotting" approaches to research in terms of how we select research problems, how we theorize, and how we view "contributions." (In short, the idea of "gap spotting" builds on the idea that there is a lack of research in a particular area, and that the identification of this "gap" can then be used as motivation for doing that particular research, and accordingly, a contribution is something that reduces this gap.)

However, the different risks associated with the gap spotting approach to research are quite many, and quite problematic. We risk going in different directions (as a research community), we risk going into the fringes (picking irrelevant research topics, and we might mistake making progress, in terms of pushing what we know even further, with just expanding our object of study (picking new topics instead of going more in-depth – to verify, or rethink.

Of course, we also need to move forward in terms of expanding and moving into new terrains, but as a research community we need to know why we need an expansion (if it is to challenge what we know, or if it is because we know enough about the things we know already). In short, staying put and dig deeper, or working more horizontally to address new topics is a central concern for us, and it is fundamentally intertwined with the temporalities of how we do design research.

Things to think about:

- Why we need to avoid "gap spotting" – the fringes (important progress/research breakthroughs might not only be in unexplored territories)

- Is it only about "moving forward" – about pushing the envelope, and about moving into new terrains, or should there also be a concern for rethinking what's been done?

- Is there a difference between "breakthroughs" and "progress" in design research?

NOW – WHAT DOES "RHYTHM" REFER TO IN DESIGN RESEARCH?

To say that something has a certain "rhythm" speaks not only to a particular pace, but more generally to a *periodicity*, and to the *regular recurrence* of something. If we think about design research processes, we can see the periods that set the overarching frames for these rhythms. There are the periods of explorations, the periods of ideations, the periods of generating design alternatives, the periods of iterations, the periods of realizations – including making and breaking, and the periods of validation (including user studies, analysis of environmental footprints, critical reflections about how the design changes practice, and reflections on how it might privilege some groups, while rendering others invisible.

At the same time, research also has its own pace. Research problems need to be understood, related research needs to be covered, appropriate studies need to be conducted, results need to be analyzed, and results need to be published. This is a different cycle than the cycle of design, but in design research these two need to come together – and accordingly, the planning and the process of design research needs to consider the rhythms of design and paces of research.

As my elaboration on the "Double Diamond" model has illustrated there is not just one steady pace throughout any design research project. On the contrary there are many different rhythms, paces, iterations, delays and other temporalities at play during a design research project. Typically, there are also not just one pace, but multiple paces that drives or slows down a design research process. Some of these are due to internal factors (for instance that a prototype needs to get finished before a user test), and some other are related to external factors (including for instance changes in deadlines, new demands on the project, changes in practice, etc.).

In fact, multiple factors at play can influence and affect the pace of design research. For instance, it might take time to find an import topic to explore, and it might take even longer to figure out how to address it. It might take time to define a design project that not only sets out to address a particular problem, but to also do it in a way that adds to an existing body of research. This contribution to an existing body of research is also something that cannot be decided upon toward the end of a project – when the results are in. Instead, this needs to be envisioned before the project has started. In fact, it is equally important to imagine what kind of research contribution the project might be able to deliver, as to image its outcome in terms of design. To figure this out, how to contribute to the particular problem, and through such acts also contribute to a more general body of research takes time.[3]

There are also a set of practical issues that takes time – and in terms of *periodicity* it takes time no matter how many design research projects you do. These practical issues are related to establishing contacts to ensure good access to relevant data, and to establish trust in relation to the stakeholders you address through your design research efforts. These efforts and processes of gaining access and building trust takes time, and this is all part of the natural pace of design research.

There are also a set of issues that are less predictable in terms of how much time it takes. Still, these issues are also recurring in almost every design process, and accordingly these issues can be said to affect the overarching rhythm and pace of a design research project. These issues have to do with the relation between the foreseeable and what is possible to estimate on the one hand, and breakdowns and breakthroughs on the other hand.

When planning a design research project these two temporalities are equally important to take into account. The foreseeable things can for instance be planned in relation to the "Double Diamond" model – time needed to be allocated for the basic activities of pre-studies, ideation, sketching, prototyping, user testing, etc. Such activities are essential to any design project and can be estimated in time based on experiences from similar previous projects.

But there are also more unforeseeable aspects that heavily influences a design research project. There might be breakdowns in communication, in the generation of design ideas, or there might be unforeseeable technical challenges. There might also be competing design research that publish new research during your own project. Such changes in the research landscape might imply that you need to rethink and re-orient your project as to ensure that it will still serve as a contribution to the stand of research you are targeting. This affects the pace and duration of the project.

On the other hand, there are also unforeseeable dimensions at play that might speed things up. In any research discipline you find the notion of "breakthrough." This is a moment that cannot be planned. It might not happen, but if and when it does it has a significant impact, not only for the results, but for the whole field of research. Such breakthroughs also, most typically, has a direct effect on the plans and estimated time frames in a project, and accordingly, it has a direct impact on the duration, the pace, and the overall planning of a project.

Breakthroughs might for instance be related to arriving at expected results faster than estimated, or arriving at new, interesting, surprising, alternative, and unforeseen results. Such results do of course affect the overall pace and duration of the research process, and it might also imply turns in the planning toward new or alternative research goals. Breakthroughs can also be related to new ideas, or the trying out of new methods approaches, or ways of seeing (theories), that fundamentally changes how the rest of the project is understood, approached and communicated. In these cases, a breakthrough instantly moves the whole project to an alternative future state. Such breakthroughs can be thought of in terms of temporal research leaps. In a sense, research breakthroughs collapse time. In a breakthrough moment the beginning and the end is merged, and it is possible to see how the initial research problem is linked to future solutions, maybe how the small-scale design experiment is connected to more full-scale solutions and implementations, or how the use of a particular solution might work in a wider set of application domains.

It is about the connection between the specific and the general – where the "ultimate particular" (Nelson & Stolterman, 2014) in the design research project is linked to larger, more abstract and overarching research goals.

THE RHYTHM OF REFINEMENTS

The typical *rhythm of design* is to work back and forth, rather than working in a linear way. It include working on different ideas, solutions, refinements of ideas, and the testing out of ideas – on pieces of papers, when working with sketches, or through the construction of research prototypes when aiming for more high-res elaborations of what a design idea might look like in practice. It is about trying different methods, materials, and techniques, and it is about going back to earlier versions and previous stages in the process. This rhythm of going back and forth is typically referred to as an iterative process (see e.g., Nielsen, 1993) – a process that is about working back and forth, over time, between ideas and materials that can manifest these ideas[4] and in that sense the overall rhythm of design has by many (see for instance the work by Donald Schön, 2017) been referred to as an iterative process, a non-linear process or a highly dynamic process.[5] This is about the rhythms of design.

In similar ways, *the paces of research* have similar recurring patterns, although here it can be more repetitive, and less predictable in terms of when (if ever) there will be any significant findings. The pace has to do with the speed of the process, and it might be about collecting and analyzing data, revisit the data, and collect even more data. It can also include things like going back to the initial research question, see if the analysis is correct, and keep looking. This is a process that takes time, and it is hard to know when any important discoveries will be made. This is the general pace of research, and why research takes time and why the duration of a research project might be so hard to predict.[6] In addition to this, there might be other uncertainties at play – for instance, do we have the right data for this research? Are we approaching it with the most suitable method? And how are our ways of seeing affecting what we focus on, and what we might see in this data?

In short, research takes time and neither research nor design is something that can be fully predicted; it can accordingly not be perfectly planned, scheduled, and conducted along a predefined timeline. In design research these two – design and research – comes together in processes where neither the design process nor the research process is linear. On the contrary, it is highly dynamic. Some days the whole project can quickly move forward. Other days might be full of surprises, breakdowns, and deviations from the original plan. This is about the temporalities of design research, and what it looks like in practice.

ON THE RHYTHMS OF REFINEMENTS AND DESIGN ITERATIONS

Usually when we talk about design processes we talk a lot about iterations and iterative processes. We suggest that design is a process of working back and forth between ideas and material manifestations of ideas. In these processes we use different materials, we sketch and we build low-fi and hi-fi prototypes. We try things out, we test, we do, we redo, and we improve. Incrementally, but steadily, we move toward the end result.

In a sense, all of these iterations can be seen as an ongoing act of polishing the design, and making the design slightly better through each iteration. As such, "iterative design" is a process of refining the design, or as formulated by M Cobanli, the founder of OMC Design Studios – "Great design is the iteration of good design."[7]

FROM REFINEMENTS TO DEVIATIONS – NEW PATHS THAT MIGHT STOP OR PROLONG THE WHOLE DESIGN PROCESS

Iterations are not only for the purpose of refinements, that is, it is not only about moving from one early prototype to a more refined version of the same thing. Instead, the iterations are also for the purpose of trying out alternatives – and by doing so it can also lead the project into new, unforeseen paths, to alternative solutions, and toward alternative end results. When we think about iterations along these lines we can think about iterations not only in terms of refinements oriented toward a polished ultimate design, but as trial-and-error processes. That is, through an iteration we try out what works, we benchmark our thinking through design, and we also notice the things that do not work.

Through these trial-and-error processes we test how our ideas work in practice, and the errors we encounter tell us that there are still things that needs to be improved, or more fundamentally challenged and changed. Accordingly, trial-and-error processes can on the one hand be seen as spiral processes of refinements, and on the other hand as cycles that move between design ideas, and the identification of problems that needs to be resolved. Of course, this has implications for just how long a design process takes. Refinements can be quick, but a continuous strive to polish a design can also be an almost never-ending story. Likewise, if the trial-and-error process constantly leads the process into new paths, it might be hard to know how to navigate these new paths to ever reach a final design.

For sure, if we go back to the double diamond model, this process looks quite straightforward – we move through two cycles of first doing divergent activities followed by convergent activities. We use the divergent phases to explore a topic, to examine different paths forward, and to ensure that we include alternatives, and we go for convergent activities when we select which path to take, when we filter out alternatives, when we prioritize, and when we decide. In reality though, this might happen through multiple cycles, and it might be hard to ensure that each iteration is convergent, as it might as well open up for new alternatives.

In the following figure, I have tried to illustrate how we can see two types of deviations here from a more straightforward, linear, and well-known process. We can call these two types "Deviation type 1" and "Deviation type 2."

> **Deviation type 1** – is "a solution in search of a problem." In this case the design process has taken some different turns, and it has over time deviated from the initial research problem. The more it deviates, the harder it will be to relate it to the initial problem, and accordingly, it might be far-fetched to contribute at a later stage (at t2 in Figure 4.5). We can call this a deviation of the design.

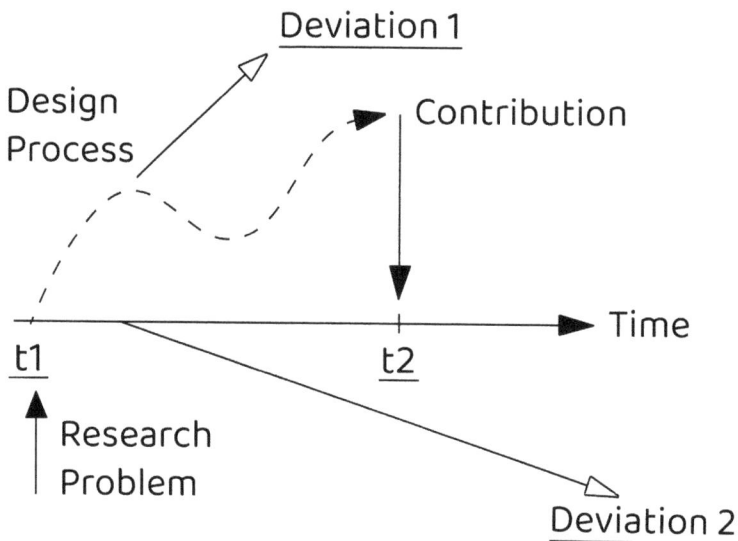

FIGURE 4.5   Two types of deviations.

Deviation type 2 – is the type of deviation that happens due to changes in practice. Although the project might follow its initial plan, there are things that have changed in practice that might imply the proposed design is less and less relevant. We can call this a deviation in practice, and again, it might be far-fetched to contribute with one such design that tries to solve a problem in a practice that has already been resolved, or it can also be the case that practice has already shifted to something else.

So how can we deal with these deviations? For sure, iterations and deviations are part of the design process, so we cannot eliminate these elements. If we tried to do so, it would probably change the process into something different than design. So, if we cannot eliminate it, then maybe we can approach it differently? In this book, I discuss such tactics. I do this by examining how changes in scope, approach, and material can allow for deviations while ensuring timely contributions. A strategy that seeks to eliminate deviations might limit the design, and a strategy that seeks to shorten the design process – to make less room for deviations over time in the first place – might also limit the level of ambition of the design research project. Related to this, I propose that design research should not only adjust itself in relation to the turns the design process takes (deviation type 1) or changes in practice (deviation type 2), but it should ultimately, and always, strive for the making of practice through design, i.e., to abandon the idea of seeing design and practice as two different things, that is to abandon the idea of a design as a solution offered to practice. Instead, I propose seeing these two as one – as designs that shape practice, as practice-making through design. In this book, I discuss how one such approach is an approach of emancipation and responsibilities, and how this is an approach to future making while staying grounded in the present – in terms of underpinning values, concepts, and hopes.

SUMMARIZING – AND MOVING FORWARD

To summarize, I have in this chapter looked at the complex interplay of fast and slow approaches to design research. I have looked at the importance of each approach, and I have discussed how the two play out and comes together in practice. I have also looked at how divergent and convergent processes of design are modeled, and I have discussed iterations and deviations in relation to these phases, and in relation to design research.

Finally, I have, toward the end of this chapter, discussed alternatives to the elimination of these deviations.

Still, there are two clear approaches to design research at the current moment – including "fast" approaches to design research that are to a large extent about applied design research, and about finding solutions to pressing problems, and there are "slow" approaches that seek to do more basic or fundamental design research with a clear ambition to make more long-term contributions. Over the next two chapters, I examine these two approaches in more detail.

Accordingly, I will in the next chapter, start with a focus on how to approach a design research problem, while aiming for a solid research contribution. In short, the next chapter, devoted to "aiming and approaches" is about approaching and tackling a contemporary research problem, while aiming for a timely research contribution, followed by Chapter 6 which is devoted to having a "focus and framing."

NOTES

1. Sometimes this is referred to as RtD – "Research-through-Design" (there are a lot of papers published on this approach, see for instance the work by Zimmerman et al., 2007).
2. In this context, AI is sometimes described as one such new design material (see e.g., Alavi et al., 2019).
3. As formulated by Sørensen (1994) in the classic paper "This is not an article" – "to identify a research topic takes time" and "you need a good excuse to take a stand" (related to an established body of related research).
4. A perspective developed by Donald Schön (2017) in his book *The Reflective Practitioner* in which he discusses this dialogue form of interplay between a designer and his/her materials at hand.
5. See for instance the book *Thoughtful Interaction Design* by Lowgren and Stolterman (2004).
6. As suggested by Einstein, "If we knew what it was we were doing, it would not be called research, would it?"
7. https://www.interaction-design.org/literature/article/design-iteration -brings-powerful-results-so-do-it-again-designer?fbclid=IwAR0XnBFIDm _vPOx3h3SXIpGlbo-8OOl9YWjtyDuOjQDfA5O-68lSNC0tAgY

CHAPTER 5

Aiming and Approaching – On Goals and Methods

INTRODUCTION

There are different ways of approaching a research topic. I decided to name this chapter "aiming and approaching," because ultimately it is about deciding what to aim for, and how to approach it through methods and studies. The aim can be motivated in terms of how the research problem is articulated, what the ultimate research goal is, the scale of the research problem at hand, or how urgent it is to address the research problem. Some problems are small and well-defined, whereas others are huge and may be hard to identify and define. For instance, many global challenges are huge, but these are often also intertwined with each other (health and clean water, industry production and food, transportation and emissions, etc). Accordingly, these research problems are big in terms of scale, and these problems also need to be further analyzed since they are intertwined with other problems and the side effects of other possible and proposed solutions.

Overall, this chapter is devoted to this important challenge of aiming and approaching. Here, aiming has to do with which research problem to address and how it is formulated, and approaching has to do with methods – through which approach, through which steps, are you approaching your aim? These questions are important to ask since there are so many options for how to formulate a research aim. But as you engage with this challenge,

DOI: 10.1201/9781003343745-5

you should always remember that how you formulate the research aim and what method you select are always tightly interconnected – *you can aim for the stars, but if you don't have a rocket you will never reach the stars.* In short, be reflective and logical when you select what to aim for, and in your motivation and selection of which method(s) you intend to apply to reach your overarching aim. Further on, you should remember that the formulation of the research aim and the corresponding research question will also be directly linked to what you can say at the end of your research process, that is, your answer to your research question. And also, it is both linked to your method, as well as directly linked to what kind of research contribution you can make – and accordingly, it is ultimately linked to what impact you will have in your area of research.

As you move from an initial idea about what to study, to the formulation of a research aim, you should know that you can decide not only to pick an urgent research problem, but you can also go for urgent research problems, that might at the same time be somehow eternal. We do have urgent sustainability problems (for instance, climate change, etc), but sustainability and sustainable lifestyles might also be an eternal challenge. We will probably not just find a solution to sustainability and then leave that question behind. Instead, sustainability will probably always be there as an area of concern. Accordingly, I will in this chapter cover the identification of "timely/timeless" research topics and ways of aiming for and approaching these topics from the perspectives of *scale* and level of *urgency*.

If a research problem is identified as addressable (being at a reasonable scale), and if it is urgent, then the research might be *timely*. If the problem is at a reasonable scale and recurring, or something that will never go away, it is probably a timeless research topic – relevant now, but also in the foreseeable future. In this chapter, I present a small model and approaches for doing such identifications. I also discuss issues related to finding a timely/timeless research topic, and I discuss how to choose a suitable method and approach (fast or slow, or a combination of these two) for addressing an identified research problem. Finally, I also discuss aspects of finding your own rhythm as a design researcher when you set out to work on a particular research problem or topic.

On an overarching level, I suggest that when picking a research topic, it is important to consider both timely and timeless issues. You need to go for something relevant (timely), but also choose a topic that belongs to an existing body of research, a strand of related work, within a particular research community. You can of course pick something completely new, but then you will have a hard time justifying the topic and positioning it, and you might find it hard to know to which body of research you should contribute. Accordingly, picking a good research topic is a matter of knowing a problem domain, and knowing a research community from the perspective of published work.

Slower movements, or more radical shifts, in our surrounding society, can call for attention, or raise awareness, around important issues and provide opportunities for ways forward. Such "timely topics" may include the latest developments in technology, such as AI, the third wave of AI, and the societal deployment of new technologies (e.g., the use of ChatGPT by OpenAI in various societal contexts). These topics can be relevant in the short-term, but these topics may not necessarily have long-lasting academic value. On the other hand, if you can choose something that is a combination of something very popular and timely (e.g., AI) and something more long-term (e.g., "the future of human cognition") you might have a good timely case that also puts you in a good position for making more long-term impact. When I say long-lasting here I am thinking about for how long it will be relevant as related work for someone else. For instance, the future of human cognition is a topic that almost by its formulation suggests that it is relevant over a period of time. In short, this illustrates the importance of finding something that is both "timely" and "timeless."

Slower movements and timeless topics can provide a deeper understanding of complex issues. For instance, to "critically analyze autonomous materialities" (Wiberg, 2023) might provide insights valuable for understanding autonomous vehicles, or studies on the intersection of AI and UX, and how AI might lead to "automation of interaction" (Wiberg & Stolterman, 2023). It might also fuel discussions on how to design for good user experiences (UX) while making use of AI as part of the solution. Such combinations illustrate the importance of working across the short-term vs. long-term research problem divide, and, if managing to shift from something that is just timely and urgent to something more long-lasting and long-term it can help in the processes of theorizing something that happens now, and can provide a foundation for future research.

In fact, I think that those who can manage to aim his/her research so that it connects a timely topic to something more long-term is something I would call "higher-purpose" design research. That is when it is not only about understanding, addressing, or solving a particular research problem at a particular moment in time, but where it can also generate new knowledge and new ways of approaching, seeing, understanding, or analyzing the problem. This would then be about an approach that is not only practice-oriented, but also theory-oriented. I would suggest that such higher-purpose design research is long-term and value-driven and focuses on important societal issues, including topics such as ethics, gender and body, sustainability, climate, and race. This type of research can be both timely and timeless, as it addresses important issues that are relevant now and will continue to be relevant in the future.

When planning a research project, it is important to start with questions concerning what to aim for, and the desired methodological approach. This allows for the formulation of a clear goal and overarching approach for the research project, and it helps to align the research problem with the formulation of a research question that is well-aligned with the chosen methodology. Once the aim and approach (method) are selected and defined, it is important to plan the research process and consider the necessary time and resources needed to carry out the project. However, it is also important to be flexible and adjust the plan as needed, based on the evolving field and ongoing conversations in the particular area of research. Still, without a clear understanding of what the overarching aim might be it is hard to move from a particular study, to more fundamental/groundbreaking implications, and to do this on time, or before it is too late. Here, there are two timelines to consider. Both how the research problem develops (typically an empirical concern), as well as how the research front develops (typically in terms of conceptual developments and processes of theorizing).

Overall, the planning of a research project is accordingly a process of relating the intended contribution to existing knowledge, anchoring it in an existing body of related research, and contributing by adding new insights or perspectives, or through a process of rethinking things we think we know, or things we have taken for granted in previous research. By carefully considering the aim and the intended contribution of the project, researchers can plan and conduct research that is relevant and meaningful, and hopefully be able to make timely research contributions.

ON SCALE AND URGENCY – PICKING TIMELY/ TIMELESS DESIGN RESEARCH TOPICS

Selecting a research topic is always central in any research process, and it requires careful consideration. At the heart of this decision-making process lies the fundamental question: *how do you pick a timely or timeless research topic?* To deal with this question, it is essential to assess two key dimensions: *the urgency* of the research problem, and *the scale* of the problem at hand.

Firstly, consider the *urgency* of the research problem you want to target. In a fast-paced world where demands for immediate solutions often overshadow the value of contemplation and depth, it is crucial to pause and reflect on the temporal dynamics at play. Ask yourself: – "Do you have the luxury to go slow?" in this particular project. Some research problems demand swift action and real-time responses to pressing challenges, while others unfold at a more leisurely pace, inviting deeper exploration and sustained inquiry. Understanding the urgency of the research problem will help you determine the appropriate tempo and length for your research process – whether it calls for a rapid response or a more deliberate, reflective, or maybe iterative approach.

Secondly, evaluate the *scale* of the problem. Maybe it can even be measured or estimated? Research problems vary in their scope and magnitude, ranging from localized issues with limited impact to global challenges that stretch across boundaries and disciplines. Consider the size and complexity of the problem you wish to address. How big is the problem? Does it align with or encompass broader societal, environmental, political, or technological dimensions? By assessing the scale of the research problem, you can estimate the potential significance and relevance of your research, identifying opportunities to make meaningful contributions to the advancement of knowledge and practice.

In exploring this intersection of *urgency* and *scale*, it is important to notice that there are some crossroads to navigate, including if you should seek to address *timely* issues that demand immediate attention, or if you prefer delving into more *timeless* questions that transcend temporal constraints. By aligning your research interests along the dimensions of the urgency and scale of the problem, you can decide on a path for your research, and decide if your focus should be on the complexities of contemporary challenges, or if maybe it is more about laying the foundational groundwork for enduring insights and long-lasting impact.

Across the following sections, I introduce a wo-dimensional matrix that integrates the dimensions of urgency and scale, providing a framework for selecting research topics that are timely or timeless. This matrix might serve as a guide for researchers seeking to navigate the dynamic landscape of academic inquiry, offering an overview, and systematic approach to identifying research topics that resonate with the pressing needs of the

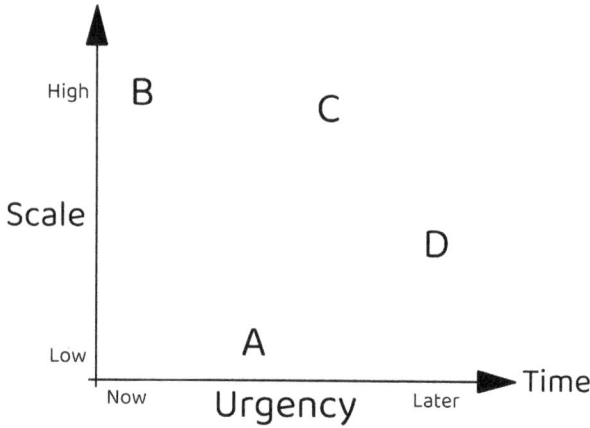

FIGURE 5.1   The scale and urgency research matrix.

present and the more enduring questions that might also be relevant in the future.

If we look at this matrix in Figure 5.1 we can see different positions here (A, B, C, D). In the following sections, I discuss these positions and exemplify what it means to aim and approach a research problem along these two scales of scale and urgency. For the urgency timeline, a position here indicates a deadline for when a solution is needed. The more to the left on this axis, the closer it is to this deadline.

Different Research Problem Positions in the Matrix

Position A – The problem is not so urgent, and it is small. This can probably be addressed in a typical design research project. The problem is well-defined, and a solution to the problem can likely be formulated, crafted, and developed within a reasonable time.

Position B – The scale of the problem is huge, and there is very little, or no time at all. This was very much the case during COVID-19, the most recent global pandemic. The problem was everywhere, on a global level, in every sector of every country. Solutions were desperately needed – from finding out how the virus was spreading, to finding vaccines, to moving work online, etc. In short, it was very urgent, and at a global scale.

Position C – Here we have all of the global challenges identified by the UN, including sustainability, starvation, education, clean water, etc.

We have global challenges, but we still have time to start addressing them. For instance, when we aim for the future by moving through green transition projects, and through the development of new technologies, etc., we can steadily move toward a more sustainable future.

Position D – Finally, we have slow-moving transitions in society that are big, but not equally urgent. It is more about the slow changes that define the times we are living in. The era of industrialization can be used as an example here. If adding additional examples, we can see how we have most recently first been through a pandemic with its related challenges, and how we are now already in a new phase in society with lots of tensions on a geopolitical level. Another example here is the fast deployment of AI at every level of our society – from how we as individuals search for information, to how digital platforms increasingly make use of AI to filter and find patterns in data. These shifts govern calls for research funding, themes for conferences, etc., and accordingly, these larger and slower shifts have a direct effect on the shorter and smaller research projects we conduct.

We can of course for a particular research project also identify several positions here. For instance, if doing sustainability research, there might be a very urgent problem (climate crisis), but to address it, it needs to be divided into a larger number of small-scale projects.

Now, let's see how we can develop this model a step further. For instance, by looking into a set of areas in this model, where we can identify areas for standard research, targeted and flexible research, and comprehensive and accelerated research as follows (Figure 5.2):

Here, this graphic model illustrates how to approach the planning of research based on the size of the research problem and its urgency. It divides the previous matrix into a model with three areas and categories of research:

A) Applied, Comprehensive and Accelerated Research – For large-scale research problems that also require immediate attention (now) and action (the upper left quadrant of the model), a comprehensive and accelerated strategy is needed. This approach involves mobilizing more resources, perhaps working along parallel research tracks, or in large-scale research programs, and using fast methods to cover

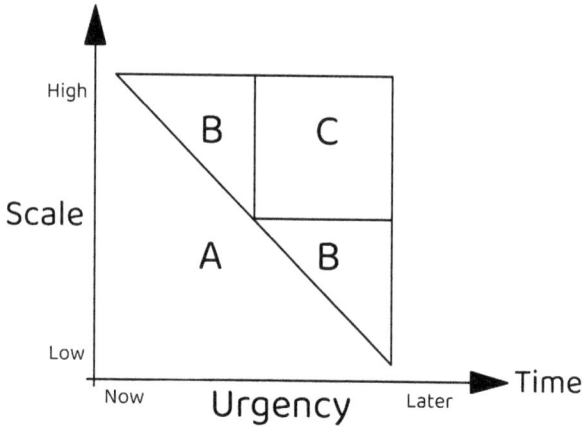

FIGURE 5.2   On scale and urgency of research problems.

the problem's breadth and depth urgently. For instance, during the COVID-19 pandemic or when the scale of the problem is lower, and there is still lots of time (the lower right quadrant), different applied projects can be carried out to move toward a long-term research goal

B) Targeted and Flexible Research – This is for scenarios when either the research problem size is significant or the urgency is rather low, but not both (the upper left and lower right areas of the model). This research strategy should include a more focused and flexible planning and approach. It implies prioritizing key areas of the research problem that are critical and can be addressed quickly. For instance, if focusing on targeted research to address the scale of a research problem, or if working in a very flexible way, and maybe include the development of new methods and approaches, for working slowly and steadily on a long-term research topic when the problem scale is more moderate

C) Basic Research – This corner of the model is about scenarios where the research problem size is large, but there is no real urgency – the need for solutions is not pressing. Here researchers can apply a traditional, and maybe even basic research approach, where there's sufficient time to explore and understand the problem in depth, and explore various solutions to the research problem. The scale of the problem motivates large-scale research funding, but since it is not urgent it can be carried out in long-term, large-scale research programs where researchers develop new methods, models, and theories and where existing solutions are evaluated, and challenged

This model can serve as a general guideline for allocating resources and defining the research methodology based on the problem's characteristics. We can also think about some simple examples of interaction design research that might fit the different areas of this model. For instance, the following examples might work as an illustration of research problems at different scales and different levels of urgency:

Example 1 – Applied, Comprehensive, and Accelerated Research

As our first example, we can think of something that could fit in area A of this model. For instance, the development of a new type of interactive system for autonomous vehicles. One such research project could aim to create novel user interfaces that allow smooth and intuitive interaction between humans and self-driving cars, including everything from vehicle control interfaces (e.g., TOR – Take-over-requests) (Bley et al., 2023) to pedestrian communication systems. Here, either the problem size or the level of urgency might be high, as autonomous vehicle technology is rapidly developing and being deployed in society, and there is a pressing need for effective interaction design solutions to ensure safe and efficient use. Therefore, the research group must quickly conduct comprehensive studies, prototyping, and user testing to develop and implement solutions. At the same time, these different applied projects might form a body of related work steadily moving forward.

Example 2 – Targeted and Flexible Research

Here we can think about an example where the problem size is large but not so urgent. For instance, the creation of a universal design standard for smart home devices. Here the goal is to create a design that is intuitive and easy to use for a broad user group, including the elderly and people with disabilities. The research problem is large, as it encompasses many different devices and use cases, but it's not immediately urgent. Researchers might choose to initially focus on the most commonly used devices or those in greatest need of improvement, and then gradually expand the research through a step-by-step process.

On the other hand, when the research problem is urgent but not large, we can, for instance, think about a project-oriented toward a rapid usability evaluation of a disaster response app. This would be a research project that quickly investigates and improves the usability and accessibility of a mobile app designed to help people navigate during natural disasters. Here, the problem is specific and urgent, as the app needs to be ready to be used effectively in a crisis. Accordingly, if in a hurry, the research group needs to act quickly to identify and resolve any usability issues.

Example 3 – Basic Research

Here we can think about a research project with the aim of developing a new set of accessibility guidelines for websites. This could be a research project aimed at evaluating and improving the accessibility of websites for individuals with various disabilities. Here, the urgency is low to moderate, as it's an ongoing effort within web design, and the problem size is manageable. The research process can afford to conduct user testing, gather feedback, and gradually devise guidelines that can be implemented over time. And beyond this, this slow process can include basic research questions such as in this case, what is accessibility? For whom? What does it acknowledge? And who is included here? And even further, how can it then be defined? And how can it be further studied and measured?

On an overarching level, this exercise to identify the scale and level of urgency is not only a matter of planning the research project, and as such also about the length and included activities in this project, but it is also a research strategic issue. Accordingly, the next section is devoted to thinking about this in terms of how it is related to such strategic issues.

AIMING AND APPROACHING – RESEARCH STRATEGIC ISSUES RELATED TO METHODS

In this section I explore the process of aiming and approaching as a research strategic issue, highlighting the importance of selecting *doable* research topics that align with the scale, urgency, and feasibility of inquiry.

So, what do I mean when I say "doable"? Well, central to the selection of a research topic is the notion of appropriateness – finding a problem, method, and phenomenon that is well-suited to the researcher's capabilities, resources, time, interests, and aspirations. An appropriate research topic aligns with the researcher's expertise and interests, offering a manageable scope that can be feasibly explored within the constraints of time, resources, and knowledge. Moreover, appropriateness extends beyond mere manageability, as it also has to do with the notion of timeliness – a "window of opportunity" where the research topic holds relevance and significance within the broader scholarly discourse in a research community. In short, it is of strategic importance to make sure that this research interest is also well-aligned with an ongoing scholarly discussion in a particular (sub)field of research.

Ultimately, the process of selecting a research topic is one such strategic endeavor – one that involves careful consideration of the *scale*, *urgency*, and *feasibility* of inquiry. In short, it is a process of examining if the research is "doable" – not just from the individual researchers point of view, but also if this view and this individual interest is aligned with a broader interest in this topic in the research community. By aiming for research topics that are both intellectually compelling and practically feasible, researchers can approach their scholarly ambitions with clarity, purpose, and intentionality, and ensure that their research efforts contribute meaningfully to the advancement of knowledge and practice in their particular field research.

So, given that you have found this balance in terms of *scale*, *urgency*, and *feasibility*, then how do you move from this strategic concern to the practical selection of a research method that can guide your research process, from the initial research aim, to results that can constitute a research contribution?

CHOOSING THE RIGHT METHOD – COMPARING AND COMBINING FAST AND SLOW APPROACHES TO DESIGN RESEARCH

The choice of research method plays a central role in shaping the outcomes of a study. From the perspective of "aiming and approaching" the selection of a research method is an iterative process of moving back and forth between the formulated research aim and how that aim can be approached through a methodological process.

Here, two contrasting approaches – fast and slow – offer two distinct avenues and paces for inquiry, each with its own set of advantages and considerations. In this discussion, we will explore the key differences between the fast and slow approaches, followed by an examination of how these two approaches can be combined to enhance the richness and depth of design research outcomes. In short, if we want to do "higher-purpose" design research, then how can we combine the fast and slow approaches in our research design? And further on, what might a combination imply for the pace and length of a research project?

Comparing Fast and Slow – Understanding the
Dichotomy in Terms of Approach

Central to a comparison of the fast and slow approaches to design research
is a fundamental tension between *immediacy* and *depth*. The fast approach
is characterized by its agility and efficiency, focusing on rapid data collection and analysis to generate actionable insights within tight time frames.
For instance, by working along the method of "quick and dirty" or "short-
term ethnography" (Pink & Morgan, 2013). In contrast, the slow approach
embraces a more reflective or contemplative tempo, allowing researchers
to immerse themselves in the nuances of user behaviors over extended
periods, uncovering deeper motivations and contextual insights.

One of the primary distinctions between the fast and the slow approach
lies in their respective time frames. While the fast approach prioritizes
speed and *expediency*, the slow approach embraces a longer-term perspective, recognizing the value of sustained engagement and in-depth exploration. This temporal dichotomy extends to the depth of insight provided
by each approach, with the fast approach often offering a broad overview
of empirical data (e.g., user behaviors), while the slow approach enables
researchers to delve into the underlying motivations and complexities that
shape those behaviors, to theorize, and to develop concepts, theories and
new ways of seeing things. For sure, this is of strategic importance to consider when formulating the aim of the research process and selecting the
research method. If aiming for a long-term process, then a longitudinal
method might be the right choice, but if going for a rapid process, then
that also comes with implications for what methods that might fit one
such time frame.

Moreover, and if expanding along the "approaching" perspective we
should notice how differences in data collection methods and analysis
techniques further distinguish the fast and slow research approaches. The
fast approach typically relies on small-scale qualitative studies (e.g., a limited set of interviews, an experiment, or a survey), or quantitative methods such as large-scale surveys, database studies, and A/B testing, that
generate data that is readily quantifiable and easy to interpret. In contrast,
the slow approach favors qualitative empirical methods such as ethnographic field studies and longitudinal case studies, allowing researchers to
uncover nuanced themes and emerging patterns that may not be apparent
through quantitative analysis or small-scale and short qualitative studies. The differences between the fast and slow approaches also influence

their perceived actionability and credibility. While the fast approach may provide immediate insights that are readily actionable by stakeholders, the slow approach offers a more nuanced understanding that may be perceived as more credible and rigorous due to its depth and thoroughness, but at the same time less "actionable."

If going for a combination of these approaches it is important not only to consider the nature of the fast approach vs. the slow research approach, but also consider the different time frames that come with each approach. In the following section, we take a closer look at what kind of synergies there might be if combining the fast and slow approach.

Combining Fast and Slow – Exploring the Synergies from the Perspective of Approaching

If thinking about combining the two approaches, we should notice that the fast and slow approaches don't have to conflict with one another, even though they can seem to be essentially different. To enhance the research process and its results, these approaches can actually be integrated in a number of different ways. Combining the fast and slow approach can be accomplished in part by employing the quick approach's agility to spot trends and patterns that the slow approach can then investigate further. Researchers can obtain basic insights that provide the basis for more in-depth investigation through longitudinal analyses or ethnographic field studies by rapidly gathering data from a large number of users. On the other hand, the slow technique can provide the quick approach's findings with more depth and perspective.

Researchers interested in combining these two approaches can enhance the understanding obtained from quantitative data by identifying the social, cultural, political, and environmental elements that impact user interactions and preferences by spending time in the natural contexts of users and tracking their behaviors over time. Moreover, merging data from both methods – through integration or triangulation – can produce a more thorough comprehension of the study subject, and through one such approach combine both the depth of qualitative insights and the breadth of quantitative patterns. Furthermore, using a mixed-methods approach – which incorporates aspects of both slow and rapid methodologies – can provide a structured framework for combining various analysis approaches and data sources. Researchers seeking this combination can then optimize each approach's advantages while reducing its drawbacks by methodically integrating quantitative and qualitative – fast and slow – methodologies.

Ultimately, the research aim, research objectives, available resources, and project constraints should inform the decision about whether to use a fast or slow approach, as well as decisions on how to combine them. Researchers can optimize the richness and relevance of their design research outcomes by customizing their methodological approach to the unique requirements of the research question and context by carefully weighing the temporal dynamics, depth of insight, actionability, and credibility of each approach.

The "scale and urgency" matrix I have presented in this chapter provides a useful framework for matching methodological decisions to the specifics of the research topic when it comes to choosing research methodologies for design inquiry. Researchers can better traverse the intricacies of design research by integrating the dimensions of *scale* and *urgency* with the distinctions between fast and slow approaches. This will help to ensure that their methodological options are tailored to the specific requirements of the research setting.

Further on, researchers must evaluate the breadth and depth of investigation necessary to appropriately address the research problem while assessing the *scale* of the research problem, which can range from local and particular difficulties to general and global challenges. The fast approach could be preferred for large-scale issues that call for quick thinking and comprehensive insights since it enables quick data collecting and analysis to identify broad trends and patterns. In contrast, the slow approach provides a more thorough and nuanced understanding of complex, multifaceted problems with long-term ramifications, allowing researchers to explore the nuances of user behavior and contextual dynamics in detail.

Similar to this, the pace of research activity is determined by the *urgency* of the research topic, which can range from current and pressing needs to long-term considerations. The fast approach is appropriate for pressing problems that require quick action because it offers timely insights that can guide decisions and actions right away. On the other hand, the slow method enables researchers to take a more deliberate and thoughtful position when dealing with topics that have longer-term repercussions and slower temporal dynamics. This allows for deeper examination, theorizing, and iterative improvement over time.

Through the process of mapping the research problem's characteristics onto the *scale and urgency* matrix, researchers can choose the most suitable methodological technique for their particular situation. In order to uncover underlying complexities and long-term trends, for instance, research problems in the *high-urgency, high-scale* quadrant of this matrix may require a combination of fast and slow approaches. The fast approach's agility can be used to address immediate needs, while the slow approach's depth and rigor can be embraced. On the other hand, research questions in the *low-scale, low-urgency* quadrant would be more suited for a purely slow approach, which would free researchers from the constraints of urgent deadlines or broad implications and allow them to pursue their investigations slowly and thoughtfully. Researchers can make sure that their research projects are both grounded in rigorous, systematic inquiry and responsive to real-world needs by matching methodological choices with the scale and urgency of the research problem. This will maximize the relevance and impact of the design research outcomes. Overall, this illustrates that there are a multitude of options available when planning for the timing, duration, and pace of a research project. It also illustrates different personal options. Some researchers might want to work on a more fast-paced and applied project, whereas others prefer to work slowly and steadily toward a long-term research goal. This matrix illustrates that both approaches are possible. Still, it is not only about the individual researcher, but also about the research community – the collaborative and collective dimension of research.

The Temporality of Aiming and Approaching – from the Individual to the Collaborative

Starting up a research process is from an individual perspective not just an objective process. On the contrary, it is in fact an intimate undertaking that is influenced by the distinct rhythms, tastes, and inclinations of the researcher.

When we start up a new research process, it is typical that we begin by moving into the reflective domain of individual pace and tempo, where we lean toward the kinds of studies, methods, approaches, and research procedures that we as individual researchers find most effective, meaningful, and comfortable. Through this process, we establish the foundation for designing research projects that are in harmony with our innate motivations and working styles by focusing on our preferred format, temporality, and rhythm in our research process.

For sure, every researcher has a unique blend of abilities, inclinations, and characteristics that influence our choice of method of exploration and learning. Which kinds of studies are you most interested in doing? Through which methods? Do you prefer the richness and depth of ethnographic research, or are you more comfortable working quickly on quantitative surveys, rapid prototyping, and experiments? By being aware of our methodological inclinations and research preferences, we may steer our research orientation in a way that best utilizes our innate abilities and interests, resulting in a feeling of genuine purpose and alignment. This is a process that is ultimately not only about finding your "voice" as a researcher, but also about figuring out who you are, your character as a researcher.

Furthermore, the pace and rhythm you select for your research processes are not just technical choices; they also speak to the length and degree of your attention span. For what length of time can you focus on a specific research topic and remain interested and engaged with it? Do you want to work in short, concentrated periods, or do you prefer to immerse yourself in long-term, continuous inquiry? We may improve our research processes to take advantage of times when we are most productive while also cultivating the patience and persistence needed for in-depth exploration by learning to tune in to our natural rhythms and energy levels.

Research is however more than just this individual pursuit; it's also, and to a great extent, about teamwork, community, and so-called "academic citizenship." Research is a collective effort that benefits from the combination of many viewpoints, discussions, and contributions. When we move from the realm of individual efforts, interests, and pace to the cooperative setting of research collaborations and academic citizenship, we expose ourselves to a larger ecosystem of mutual assistance, teamwork, team efforts, shared responsibilities, and shared knowledge creation. Following the idea of "standing on the shoulders of giants" (Scotchmer, 1991), it is a process where we work together, build on each other's work, provide feedback and reviews, and we revise and resubmit. As Sir Isaac Newton himself acknowledged, "If I have seen far, it is by standing on the shoulders of giants."

In short, we work together in many different ways, and these collaborative elements of the process also influence the temporality of the process. We can achieve more together than if we work alone, but this feedback from our peers also takes time to process – In short, it can both speed things up and slow things down.

These research collaborations can also take many different forms, including project and research program collaborations, interdisciplinary teams, research groups, and individual research collaborations. You can get an overview of these networks by asking yourself questions like, Who are your peers and colleagues? Who serve as your role models and mentors? Who is in your team? And where are your research colleagues located? These inquiries address the interdependence of the scientific community and emphasize the value of building deep collaborations and networks – across universities, departments, institutes, and research areas – that promote cooperation, mentoring, and reciprocal development. It also provides opportunities for research exchange visits, interdisciplinary collaborations, and the formation of larger research constellations. Further on, and through these collective efforts, opinions, and contributions, we not only advance knowledge but also strengthen the connections and research collaborations that make a research community work – it is in fact this work of working together that enables research communities to exist and grow, and in return, these communities enable the individual researcher to grow!

Research is in these terms a collaborative effort that thrives on the interchange of ideas and viewpoints. As researchers, we are members of a wider community that supports our intellectual development. This community consists of mentors, peers, colleagues, and role models. We participate in academic citizenship and community service while working in teams, research groups, and projects. These collaborations foster a collaborative, knowledge-sharing, and supportive environment, no matter if you are a young scholar about to enter a new research community, or if you work as a senior researcher together with others in your community.

Planning the Research Process in Terms of Aim and Approach

Given that you are aware of, and belong to a particular research community, you also need an additional set of things to get you started. Launching a research process requires not only a research problem, and a research community. As I have pinpointed in this chapter, you also need a research aim, a methodology, and an organized approach. A thorough planning approach that includes goal formulation, technique selection, alignment with larger contexts, process planning, and recognition of the iterative nature of research is fundamental to this attempt. For the research project to be coherent, rigorous, and ultimately successful, each part of this arrangement is essential.

First of all, any research effort must define the research problem and the goal. This is an essential first step. This entails defining the parameters and structuring the research, laying the groundwork for further explorations. To do this, researchers need to go through a thorough process of defining not only the problem and the object of study, but also their goals, ideas or hypotheses, and research questions. Determining the research's significance and relevance requires clearly stating how it will contribute to an existing body of research or fill some gaps in the literature. In short, researchers establish the foundation for a targeted and intentional study with a well-defined goal, and researchers need to make sure that their goal is aligned with larger settings in addition to defining it. This requires having a thorough awareness of the background of the research problem, including its importance, urgency, and ramifications. In addition, researchers need to think about when to offer a solution or make a contribution so that it fits in with how scholarly discourse is developing. The coherence between the research objectives, the problem context, and the proposed methodological approach is facilitated by this alignment, which in return establishes the way for a thorough and significant research exploration.

Once a specific goal and alignment with wider settings have been determined, you can as a researcher move on to choose and develop the particular research methodology. This crucial step necessitates giving serious thought to a number of factors, such as the nature of the research questions, the degree of insight desired, and the resources available. Researchers can choose between mixed-method, quantitative, and qualitative methodologies, based on the particular needs of their research. Through these decisions, you create the foundation for producing significant insights and expanding scholarly understanding by carefully choosing and crafting a suitable research approach.

Once the goal, strategy, and method have been established, you as a researcher need to organize your work very carefully. This entails creating a detailed process plan for the research, including deadlines, checkpoints, and deliverables. Through rigorous planning, researchers guarantee optimal resource distribution, proficient time management, and compliance with pre-established goals and objectives. Furthermore, a well-organized research plan acts as a road map, assisting researchers in navigating the complexities of their research process, and providing them with the clarity and confidence to overcome any barriers or obstacles associated with the research process.

Lastly, researchers typically need to recognize that their approach is iterative. Research is a dynamic process that is constantly adjusted and modified in response to changing scholarly discourse, new research activities, and new discoveries. In order to keep up with advancements, researchers must be alert to changes in the area, and constantly update their ideas and methods. Researchers can guarantee the impact, validity, and relevance of their discoveries in the dynamic field of academic inquiry by adopting this iterative strategy.

In conclusion, researchers looking to carry out significant and influential work can use the above-described approach to research process planning as a foundation. In addition to this, researchers can confidently and clearly address the practicalities of the research project by carefully organizing processes, selecting appropriate methodologies, realizing the iterative nature of research, rigorously articulating goals, and carefully planning procedures. By following these guidelines, researchers can improve the rigor and coherence of their work, and make a substantial contribution to the body of knowledge in their particular domains.

To make sure that our research contributions are based on in-depth exploration and significant interaction with the academic community, we move through the research process by embracing a process of connecting to, anchoring, and adding to the ongoing discussions in the research community. By carefully and strategically organizing our research process, we open the door to important discoveries, game-changing understandings, and enduring contributions to the expansion of the existing body of research in a particular research area.

AIMING AND APPROACHING YOUR RESEARCH – BALANCING BETWEEN INDIVIDUAL AMBITIONS AND MOVEMENTS WITHIN THE COLLECTIVE RESEARCH COMMUNITY

Any individual researcher will find themselves at the intersection of personal motivation, interests, and explorations on the one hand, and research community advancements at a collective level. Further on, any individual researcher has to navigate the balance between individual agency and collective progress in their research community. For every individual researcher, the process of aiming and approaching research typically begins with an introspective process closely linked to matters such as personal passions, curiosities, and scholarly pursuits. What questions ignite my intellectual curiosity? What topics compel us to delve deeper, pushing the boundaries of knowledge and understanding? What challenges do I find challenging and inspiring? These questions prompt us to define the scope or aim and it has a direct impact on our approach, and it sets the stage for the contributions we aspire to make to the academic community and the related body of published work.

But beyond viewing scholarly inquiry as a one-person isolated endeavor that only has to do with the individual's interests and personal motivation we should acknowledge that it is deeply and fundamentally entwined with other researchers, the research community, the changing currents within the research community at large, and movements in related research fields. Scholars who seek to make timely contributions to a particular research community must stay aware of the current discussions, developments, and emerging paradigms that emerge and unfold in their fields in order to carve out a position for themselves within the larger research landscape, and they need to understand how to add to on ongoing discussion in a particular research community. For instance, in the research field of HCI – human–computer interaction, we have already been through three "waves" of development within this field of research (Bødker, 2006) and we might now start to see the contours of a 4th wave (Frauenberger, 2019) of HCI. Here, the proposal to add a "4th wave" is completely aligned with how the first three waves of development has been described (not at least by using the same vocabulary of "waves" instead of suggesting a new "phase," "shift," or "level").

Further on, researchers may make sure that their contributions are conceptually novel and forward-thinking by placing their research goals and perspectives within the broader framework of disciplinary developments, and relating their research efforts to major shifts or directions within their particular field of research. This relationship between individual research ideas and the overall development of the research community is complex and dynamic, marked by an ongoing flow and development of concepts, frameworks, and ideas. Researchers engage with existing theories, approaches, and debates to inform their own scholarly pursuits as they work to define their research agendas and trajectories, drawing inspiration from current, and emerging discourses in the research community. For instance, if you are running a research project on human–plant interaction, where you are developing an app to make people more aware of the plants we surround ourselves with in our everyday lives, you might have an interesting case and an interesting piece of technology. However, it is when you see how you might be able to contribute to say a growing strand of "human–plant interaction" research, or studies on "more-than-human worlds" that you can make a more significant contribution to a research community where these perspectives are developed.

This process of "aiming and approaching" your research toward a particular research community, and a particular conversation within a research community, is accordingly an iterative and collective process that is prone to modification and adaptation in response to changing disciplinary paradigms, intellectual currents, and emerging streams of research in the research community. Researchers need to be good academic "listeners," and they need to be flexible and adaptable when the research community changes and new findings are made; they need to modify their study objectives and frameworks to take into account new possibilities, problems, priorities, and ways of talking about things. In short, researchers need to be good at listening and communicating. A community is not a stable object. On the contrary, it is constantly formed by the discussions that constitute a research community.

In this way, the process of "aiming and framing" one's own study is closely linked to the research community's overall and collective development. This mutually beneficial relationship fosters advancement, creativity, and intellectual growth, and it fosters a sense of belonging – a way of being in the research community, being with the research community, and maybe ultimately about being someone in the research community.

FROM AIMING AND APPROACHING, TO FOCUS AND FRAMING

This chapter has focused on the methodological orientation of moving fast vs. moving slow with a particular focus on ways of "aiming and approaching." In doing so I have discussed how aiming has to do with which research problem to address and how it is formulated, and how approaching has to do with methods – the ways through which you approach your research aim.

In the next chapter, the focus will shift from these methodological concerns to ways of working with the theoretical orientation of your work. Here, "focus and framing" will be the guiding notions for the next chapter. These two notions have to do with what is in focus, and the framing has to do with through which theoretical lens the work is analyzed, understood, and presented. In short, it has to do with your object of study and your theoretical lens.

CHAPTER 6

Focus and Framing – Object of Study and Lens

INTRODUCTION

In the previous chapter, we covered aiming and approaching from a methodological viewpoint. In doing so, we looked at different approaches and suitable methods, and we discussed the importance of having an overarching aim for your research process. In this chapter, I add to this toolbox by focusing on the "focus and framing" of your research. For sure, we cannot do everything, and to make a timely contribution, it is also important to carefully select what to focus on, and how to frame it as a contribution to a particular field of research. Having a clear focus is typically about selecting and filtering out alternatives – to be selective and pick a particular object of study. Framing, on the other hand, typically has to do with the ideas and the theoretical angle of your research project. In short, how do you "see" your project? Through which conceptual or theoretical lens? And how is this way of seeing unique? Further on, you might need a "rhetorical twist" – to foreground what you see as the most interesting perspective – this is the framing that is so tightly coupled to your focus, your object of study. In this twist, where you decide what framing to go for, you open up for how you will tell your story, how you will present your main arguments, and how you articulate and position your theoretical

DOI: 10.1201/9781003343745-6

contribution. In short, this is about the core idea, or the central argument, in the research project rather than its empirical foundation or practical orientation.

If you are a PhD student aiming for a PhD in HCI or Interaction Design Research, you should know that selecting a focus and a suitable framing is in many ways a multifaceted task. It can allow for hands-on engagement with real-world issues, often referred to as "engaged scholarship," or a more reflective stance aimed at developing the theoretical frameworks that underpin the research project, sometimes described as the "think with" approach. As you begin this work, consider your position: are you inclined to make an immediate impact with your research – right away, – or do you gravitate toward crafting the conceptual tools and theories for others to apply? In a sense working along a different, maybe more long-term temporality. No matter if your research is practically or conceptually oriented, you need a focus and a framing. The way you present your research can either be very straightforward, practical, and direct, or it can be a more delicate endeavor where you might need to describe growing theoretical perspectives and how you contribute to these growing strands of work. As such, the focus and framing of your work are closely linked to your academic "voice," that is how you prefer to talk and write about your research. If this voice is linked to changes in practice, it might also be about reaching out to the surrounding society. If this voice is mainly about theoretical developments, it needs to reach out to your intellectual peers. Since these paths are so different, it will also impact the temporality of your research process.

This chapter is designed to assist you, as a researcher, in finding your voice and accordingly your pace and rhythm within the field of HCI and Interaction Design Research. Beyond the formulation of a suitable aim, a research question, and selecting an appropriate method, this chapter is also very much about you, about finding this "voice," your way of constructing a solid case, and about how to align that voice with an ongoing conversation in the field of HCI/interaction design research.

Hopefully, this chapter can serve as a guide for you in identifying your focus and a support for conceptualizing your research. As we have already seen in the introduction to this chapter, this encompasses more than just choosing methodologies or types of data; it's about defining your role in the wider academic landscape. Ultimately, the research identity you adopt and develop should shape every facet of your work, from the challenges you address to the solutions, innovations, or alternative perspectives you develop. By doing this, you can make your academic voice louder in the research community.

These two approaches, the "engaged scholarship" as the practice-oriented approach, and the "think with approach" as the theory-oriented approach, can also be thought of from the perspective of applied research and basic research – two classic approaches to research where the applied approach is typically described as being faster and more direct, whereas the basic approach might take a longer time, but might come with more groundbreaking impact. In this chapter, we will have focus and framing as our core perspective, and here I suggest that if you are leaning more toward applied or basic research, it has implications for what you will focus on and how you frame your research. And in return, these choices have implications for the temporality of your research process.

APPLIED VS. BASIC DESIGN RESEARCH

In design research, we can see two distinct paths: basic research and applied research. The former is often regarded as a more drawn-out process, typically producing outcomes that reverberate over a longer period of time. It is sometimes viewed as the "finer" approach to research, with a focus that rests on foundational questions and theoretical concerns.

In the area of HCI or Interaction Design, basic research, also known as fundamental or pure research, focuses on the foundational aspects of interaction design. It is a process geared toward knowledge generation for its own sake, rooted in curiosity and a drive to understand the underlying principles of, for instance, how users interact with technology and each other within digital environments, what "interaction" is or how it can be defined (for related work on this question, see e.g., Wiberg, 2012; Janlert & Stolterman, 2017; Bergström & Hornbæk, 2025; Wiberg, 2018), or how it should be studied and understood. Basic research in this field often appears more abstract and theory-oriented, concerned with questions that may not have immediate practical applications but are essential for the field's advancement. For instance, a basic research study might explore how human cognition and perception affect the usability of digital interfaces without a direct intent to create a specific product or suggest any implications for design. In short, there is no goal beyond generating new knowledge about these things. In such projects, HCI or Interaction Design researchers might conduct experiments on the ways visual information is processed, or how decision-making is influenced by interface design. The findings could then contribute to a theoretical framework that informs future applied research or product development.

Basic research in HCI/interaction design is primarily concerned with exploring the fundamental principles and theories behind interaction design, and ways of defining and describing the principles that govern human–computer interaction. It aims to build a deep understanding of the elemental forces at play within the field, such as cognitive processes, aesthetic principles, user behaviors, ergonomics, how to define what is a user, to what extent the notion of "user" is an accurate description of a human, and who is considered a user in the first place, as well as the cultural and social ways in which interaction design is embedded. By asking questions about why things are the way they are and investigating without the immediate need for practical application, basic research adds to the collective knowledge base of interaction design, providing insights that may not have been discovered through more applied or goal-oriented research.

Another example of basic research in the area of interaction design might involve studying how people interact with different types of content on a touchscreen. The goal of such research would be to understand the nuances of user engagement, distraction, what "touch" is as a bodily engagement with technology, and the cognitive load involved in complex interactions. The outcomes of this research could lead to the development of new theories about what it means for people to interact through "touch" that could guide future applied research on touchscreen interaction and the use of touchscreens in interaction design.

Applied research, in contrast, is directly aimed at addressing current, often urgent, challenges in practice. It's focused on creating practical solutions to identified problems and often involves the development of new technologies, systems, methods, or processes. The immediacy of applied research is its signature; it translates and applies the theories and knowledge generated by basic research into tangible outcomes that can be implemented and tested in real-world scenarios.

In interaction design, applied research would likely involve requirement analysis, user studies, prototype development, and iterative design, all directed toward understanding needs, identifying problems, and arriving at a solution that helps a particular group of people. As such, applied interaction design research zeroes in on contemporary, urgent problems – the kinds of issues that demand rapid responses and practical real-world solutions. This branch of research operates at a rapid pace, fueled by the pressing need to address real-world challenges as they unfold. Here, we find not only research topics identified by the research community but also thematic calls for research funding and political calls. In short, there are many stakeholders in society interested in addressing these societal problems, and applied research is, in this sense directed toward solving specific, practical questions and problems that have immediate and direct relevance to the surrounding society. Further on, applied HCI/interaction design research aims to create knowledge that can be applied immediately to design interactive products, systems, or services. In interaction design, applied research might involve developing and testing new interface designs for smartphones to improve accessibility for users with disabilities. It might also be about designing and evaluating a new navigation system for an e-commerce website to increase conversion rates and improve customer satisfaction. Another example of applied research within the area of interaction design research could for instance be an investigation into the effectiveness of voice-activated interfaces in car dashboards. Researchers in this scenario are seeking to solve a real-world problem – enhancing driver safety by minimizing distractions. In terms of activities and methods in the research process, they could employ user testing, prototype development, experiments, and iterative design processes to create a tangible product or solution that can be implemented in vehicles.

As a researcher in interaction design, the decision between basic and applied research pathways should align with your personal motivation, vision, and desired impact on the field. If your aim is to contribute to a broad understanding and conceptual development, basic research offers an approach to deep exploration. However, if you're driven to make an immediate difference by solving pressing problems, applied research will allow you to put your findings into action. Your skills and competencies also play a significant role in this choice. Basic research may demand a more theoretical and methodical approach, often requiring strong analytical skills to interpret data and formulate hypotheses, as well as a creative mind to formulate new concepts, models, and theories. On the other hand, applied research typically requires a practical skill set, including prototype development, user testing, and iterative design capabilities, and a different kind of creativity geared toward innovative ideas for innovations and solutions.

Ultimately, the impact you hope to achieve, in combination with your own interests and how you choose to develop your academic "voice," will guide your choice. Basic research has the potential to change how we think about interaction design and can influence the field for years to come. Applied research can change how we interact with the world around us today, solving immediate problems with innovative design solutions. Both are crucial for the field's growth, and coming back to Chapter 5, we are a research community, and it is what we do not only on an individual level but together as a research community that drives and advances our field of research. Through a combination of applied and basic research projects, we develop both conceptual lenses and practical solutions. This gives our field both rigor and relevance – right now and over time.

THE TEMPORALITY OF YOUR RESEARCH

Upon entering a PhD program, the foundational question that arises is about the direction of your research. This involves setting a clear *focus that* will influence the course of your work and, in return, the influence of your work within the academic community.

Determining the *focus* of your research is a substantial decision. It's not just about choosing which methods to use or what kind of data to collect. The focus embodies the essence of your academic endeavor – what problems you'll solve, what knowledge you'll contribute, and how your findings might be used in the future. It's about aligning your research with both your personal aspirations and the needs of the discipline. This alignment shapes your role and your "voice" in the research community and sets the parameters for your intellectual contributions. It is through your academic voice that people in the research community will get to know you – through your work, and it is through this voice you will communicate your contributions as written proposals in the academic papers you write for conferences and journals in your field.

Further on, your research *framing* is equally crucial. This encompasses the broader context of your work, such as the theoretical lens through which you view your research question, the narrative you construct around your findings, and the concepts you develop and use to discuss your work. This shapes how others perceive your academic identity and reflects the values, priorities, and ways of seeing that drive your inquiry.

As you decide what to focus on, and how to allocate your time during your PhD program, the immediacy of the research problem should be a significant factor. When faced with a pressing issue that requires a swift resolution, applied research might be the most appropriate route, allowing for practical application of your results. Conversely, if you're drawn to a question of fundamental importance – perhaps one that addresses core theoretical concerns or whose answers might inform future generations of scholarship – pursuing basic research could be more suitable. This route might not offer immediate results but can contribute to a lasting foundation of knowledge upon which others will build. On the other hand, if your attention is captured by pressing, tangible issues that require timely and concrete outcomes, applied research may be the path you choose. This path is characterized by its practical application and is directly aimed at producing solutions or improvements to current challenges within the field of interaction design. It's for those who are motivated by the potential for immediate impact – to see their research applied in real-world contexts.

For example, if you are intrigued by the challenge of improving the user experience for the visually impaired when interacting with touchscreen technology, applied research would guide you to develop prototypes and conduct user testing. The goal of one such research project could, for instance, be to create tangible, actionable solutions that can be implemented rapidly and effectively. If choosing this path for your research, you could work closely with technology developers, end users, and other stakeholders, ensuring that your research outcomes directly address the specific needs and constraints they face. This kind of research is not just about understanding the world – it's about changing it on a very concrete level. As an applied researcher (or "engaged scholar"), you would be on the front lines, translating knowledge into practice and real-world solutions. If so, your work would likely involve collaborative projects with industry, workshops with users, and iterative cycles of design and testing, all aimed at creating something usable and beneficial in the immediate future. In an applied research setting, the success of your PhD project is measured by the utility and adoption of your research findings. Your academic identity or "voice" can in such terms be tied closely to the innovations you develop, the recommendations and design implications you formulate and propose, and the real-world problems they seek to solve. Your legacy as a researcher will then be marked by the positive changes you initiate.

In a sense, your choices regarding *focus* and *framing* define not only your voice but also the contours of your academic career. These decisions are crucial as they determine not only the type of researcher you will become but also the potential impact of your work on the wider field.

Across the following sections, we will take a closer look at a set of examples that can further shed light on the importance of *focus and framing*, and how this is linked both to your own academic "voice" as well as to this choice between applied and basic research.

EXAMPLES ON FOCUS AND FRAMING

When focusing and framing your research in the area of interaction design, it's important to reflect on whether your overarching research approach is more aligned with immediate applications or fundamental exploration. Let's consider the following three examples to illustrate how focus and framing might manifest in basic and applied research contexts.

Example 1 – Sustainability in Interaction Design

If we start with this first example on sustainability in interaction design, we can think about this from both an applied perspective as well as from a basic research perspective.

In response to the urgent demands of environmental sustainability, you might concentrate on designing user interfaces or designing interactive systems that encourage sustainable behaviors. The research might involve developing and testing digital systems and tools that reduce energy consumption or analyzing user interaction data to promote greener practices and green transitions. The framing here is design and action-oriented, aiming for direct environmental impact. On the other hand, from a more theoretical viewpoint, you could also explore the principles of sustainable design from a psychological and sociological standpoint. This might involve long-term studies on how people perceive and interact with "green" technologies, their attitudes toward the UN's global sustainability goals, or the integration of sustainable practices into the area of design. Here, the framing is exploratory, providing a comprehensive understanding that underpins future research. Further, and even more fundamentally, would be to apply a critical approach and ask for whom this is sustainable, if it is sustainable for just some privileged people, a limited set of stakeholders, or if it is truly sustainable for the whole planet.

In HCI/interaction design research, there is currently a growing interest in "more-than-human" approaches to HCI and interaction design research (see e.g., Camocini & Vergani, 2021; Giaccardi et al., 2024). This perspective suggests that we need to shift our focus from human-centered to more-than-human-centered approaches where we consider other than human species, and even plants through the lens of sustainability. For sure, what might be sustainable from a human perspective (for instance, to make sure that we produce enough food on this planet for everyone to prevent starvation, which isone of the UN's sustainability goals), might not be sustainable from other more-than-human perspectives. As noticed by (Wiberg & Teigland, 2024), modern profound technologies such as electric cars get accepted and are seen as a solution to fossil-fuel-driven cars. Still, these cars come with a significant environmental footprint.

Example 2 – Artificial Intelligence in Interaction Design

A second example is research on AI – Artificial Intelligence in Interaction Design Research. If we think about the recent developments in AI, we can from an applied research perspective, go into the design of AI-driven interactive systems guided by ethical guidelines for AI interactions, and we can craft interfaces that ensure transparency and user control along the ideas of responsible, explainable, and even "graspable" AI (Ghajargar et al., 2021). Applied research might also involve developing prototypes for AI-driven user interfaces that safeguard privacy, integrity, and agency, and we can conduct user studies to ensure these principles and ethical guidelines are embedded in the design of these systems.

Alternatively, your research might focus on theorizing the potential futures currently being shaped by AI in interaction design. This could include speculative design methods to envision how human–AI relations and interactions evolve and affect societal norms. Further on, you might develop models for how to integrate interaction and automation of interaction. For one such example, see for instance the paper "Automation of interaction" (Wiberg & Stolterman, 2023). If going for this framing of your work, you would be framing your research with a long-view temporal lens, aiming to build theories that will inform future design considerations and user studies. This also then illustrates how there is no conflict or sharp divide between the applied and basic approach to interaction design research. On the contrary, the two approaches can be combined, and if done right, they can inform and scaffold each other.

Example 3 – Social Justice in Interaction Design

Finally, the third example here has to do with the emerging area of social justice research in the research domain of HCI – human–computer interaction and interaction design research (see e.g., Dombrowski et al., 2016; Bates et al., 2018; Chordia et al., 2024; Öhlund & Wiberg, 2025). If tackling social justice, in terms of seeking to bring social justice to the design and use of interactive technologies, an applied research approach could center on creating accessible and inclusive design tools or digital platforms that facilitate social change. This might include designing digital spaces that amplify marginalized voices, giving them visibility and a voice (see e.g., Nordin et al., 2023) or conducting action research alongside marginalized communities to identify and address specific design-related inequalities. On the other hand, if exploring this topic with a focus on basic research, you would lean toward the foundational side, and you might investigate the historical role of design in either perpetuating or challenging social inequities. Alternatively, you might want to go in-depth on questions concerning how digital technologies might manifest, support, or even deepen existing inequalities. This research could examine how design practices have been influenced by cultural and social constructs and how these might be deconstructed to foster justice in design outcomes. Further on, it can be research geared toward understanding the basic principles of social justice and how it informs contemporary research on social justice HCI. For instance, by identifying existing streams of research or proposing alternative paths forward, such as post-growth HCI (Sharma et al., 2023).

In each example, whether applied or basic, the *focus and framing* of your research reflect a commitment to addressing significant, timely issues in our society while ensuring that the research is anchored within ongoing HCI/interaction design research. Your chosen path – whether it leads to practical applications or theoretical explorations – will shape the contribution you make to the research community and the impact of your work on society.

DISCUSSION – FOCUSING AND FRAMING – YOUR RESEARCH PLAYGROUND

As we now reach the concluding part of this chapter, it's essential to reflect on the broader implications of your choice between applied and basic research, whether you have just started your PhD journey or if you are a more senior scholar. With a clear *aim and approach*, together with a clear focus and framing, you know where you are heading: your overarching method, the particular things and aspects you focus on, and your take on these things, that is your framing. With the aim, approach, focus, and frame in place, you can start developing your academic voice and character, and you can start being aware of how you conduct the research you do.

Still, these decisions aren't only for you; it is also about understanding your position in the larger research community and how you can best contribute to the dynamic field of interaction design research. In this final part of the chapter, I will therefore not just summarize, but also propose a set of questions to reflect upon.

How Do You Position Yourself in the Research Landscape?

Doing a PhD is a significant commitment, and as a PhD candidate, you must ascertain where you stand – whether you're the type to be at the forefront, actively engaging with immediate challenges and creating tangible outcomes (for instance, through the "engaged scholarship" approach (Mathiassen & Nielsen, 2008; McKelvey, 2006; Holland et al., 2010; Glass & Fitzgerald, 2010), or if you prefer to operate from a strategic distance, constructing the theoretical frameworks that others may rely upon (the "think with" approach). However, this choice is more than just a simple decision; it reflects your identity as a researcher and how others will understand you as a member of the research community. Being clear about this is accordingly a matter of academic citizenship – if you are clear about who you are and what you do you are also easier to approach and, accordingly, easier to work within the community. In short, are you doing more applied and practical work, or are you more geared toward theory-driven approaches? Further on, and no matter if you go for an applied or basic approach, you need to find your peers in the community. In short, who is doing something similar? And how do you add to the work done by your peers?

For Whom?

So, when you know who you are as a researcher, you then need to ask yourself the question, "for whom are you doing this research"? Research does not exist in a vacuum. It's meant to be disseminated, discussed, and built upon. Your audience – whether they are fellow academics, industry professionals, or the general public – will in part form and inform the nature of your research. Applied research often resonates more with those in the industry and immediate stakeholders who seek solutions, while basic research tends to engage an audience of theoreticians and fellow academics who value the advancement of knowledge for its own sake. Accordingly, when you position yourself as a researcher, and when you position yourself to whom you have as your intended audience, it also governs from whom you will get feedback, critique, and appreciation, where you direct your intended impact, and to what extent your research matters for someone else – no matter if this someone is within or outside academia.

How Much Time Do You Have?

Given that you have been able to position yourself and that you have identified your intended audience, your peers, and your network, you need to think about the temporalities at play here. First and foremost, if you are a PhD student, you need to spend your time wisely. The length of a PhD position is, in most cases a finite resource, and it's crucial to use this time effectively. The urgency of the research problems you're addressing might compel you to take a more applied route if immediate impact is preferred. Alternatively, if you have the luxury of time and the issues at hand demand a more in-depth approach, then going into basic research could be the right choice. Remember, the time you have is limited, and how you choose to spend it will shape not just your PhD experience but potentially your entire professional trajectory.

But this question is not only for PhD students. The question concerning how much time we have is an eternal and recurring question for everyone who wants to work as an active scholar in a research community. The community is not a static entity. Members are joining and leaving over time, and the research-oriented conversations that drive the community forward also change over time. As an active scholar, you need to be aware of these slow and ongoing conversations, you need to be able to tap into these conversations, and you need to contribute to these conversations through published papers. Accordingly, the question concerning how much time you have is a relational one – it has to do with what you can say (related to your own research) and how this adds to the ongoing conversation in your research community. Given that this conversation is constantly changing, you need to make your contributions to it before the conversation has shifted to a different focus. As such, being timely here has to do with your ability to add to this conversation in your field of research.

What Do You Want to Do with Your (Limited) Time?

Coming back to thinking about the temporalities at play if you have a PhD position, it is important to decide how to spend your time during your doctoral studies, and this is a question of priorities. If you are motivated by the potential for quick results and immediate impact, and if you prefer to work in close collaboration with stakeholders from the surrounding society, then applied research will likely be more fulfilling. However, if you're driven by curiosity and the desire to contribute to a lasting knowledge base, then basic research may be more aligned with your goals.

This question concerning the limited amount of time is again not just a concern for PhD students. An academic career is short, even if you stay in academia for decades. You will only be able to do a limited number of projects, participate in a limited number of collaborations, and be a member of a limited set of research groups, teams, institutes, and departments. In short, spend your time wisely. Nowadays, quite many scholars go on a postdoc after having successfully defended their PhD. This is a good opportunity to see a different academic environment, get involved in new research projects, and build your academic network. Where to go for your postdoc is an important decision. It is not just about going somewhere, but it is about going somewhere where they do the kind of research you prefer to do. If you work together with others who do similar things, it is easier to grow as a scholar, and you will be surrounded by people who can both inspire you and give you important feedback and support.

After your postdoc, you should continue to be selective, and you should seek to further develop your academic voice and character. Maybe you can already start to see the contours of your research path from your PhD to how you have continued during your time as a postdoc. As you continue to work, you should on the one hand focus on the research project at hand and the next steps forward, but it is equally important to also look back and see how you can *connect the dots in retrospect* in the words of Steve Jobs, founder of Apple. As the years go by, it will be increasingly important to describe what your focus has been, not only for your CV, but so that people in your research community can understand your background, your overarching topic and approach, and maybe even where you are heading – your aim.

FOCUSING AND FRAMING – AND DEVELOPING YOUR ACADEMIC VOICE

Focusing your research means narrowing down and setting a clear agenda. What problems will you tackle? What contributions will you make? What aspects or dimensions will you work on? This is your focus. Framing, though it might sound straightforward, is quite intricate. It's about how you filter out these aspects – your perspective, your priorities, your methods, your core perspective, and ultimately, this is about your voice in the field. It's about deciding who you are as a scholar and how you should communicate your research to other scholars in the field, and to the surrounding society. For sure, how you frame your research is not only an internal process for you. Beyond this, it is also a process that will influence how others see you as a scholar. Are you, for instance more interested in practical or theoretical implications? Are you more interested in small-scale details or larger patterns? The things you focus on will be the things you communicate, and accordingly, it will influence how others understand you and your work. In short, your focus has an impact on your academic voice.

In this context, your academic voice is accordingly not only about how you present your research, orally or in written form. It is also a unique package of your perspective, knowledge, experiences, and the theoretical and methodological approaches you employ. It is about the data you highlight, the arguments you present, and the conclusions you draw. Further on, it is how you treat related work from your community, and your ability to give credit to those who are doing similar work as you. Further on, your voice is about your ontological and epistemological grounding. It is about how you see the world, and your understanding of how we can gain knowledge about the world around us. Your "way of seeing" will form your way of presenting.

In addition to this, your academic voice is what distinguishes your contributions from those of others, and, at the same time, as you develop your voice, it also becomes the lens through which you approach things and the means by which your audience and your community, understands your work.

Developing your academic voice as a scholar is a dynamic and ongoing process – throughout your academic career. It is a process, or a continuum, that requires introspection, deliberate choices regarding your research focus and framing, and continuous engagement with your academic community. Your voice will evolve as you progress through your PhD and beyond, but its essence will always reflect the agenda you set, the perspective you hold, the methods you utilize, and the way you communicate – orally and in written form. Embrace this process, for it is through your unique voice that you will make a lasting impact in your field of research.

Quite many scholars think that they must develop a voice that is "academic enough" for the research community they belong to. They might learn very complicated words and they might develop a complicated language for talking about quite simple things. This is a dangerous approach, as it might lead to situations where people do not understand what you are talking about, and for those who develop this way of talking, it can almost turn into jargon where you "hide behind your own words." In fact, talking about simple things in very complicated ways might not be the best approach. Let me expand a little bit on this thing.

If we think about a 2 × 2 matrix where we have simple and complex problems on one axis, and simple and complicated writing on the other axis, we should avoid the square that is about writing about simple things in a complicated way. To write about complicated problems in a complicated way might be ok, but even better would be to write about complex matters in a simple way. Take the classic formula $E = MC^2$ as an example here. The complexity of this part of physics is captured in a simple and elegant form. For sure, this is probably one of the reasons why this formula is so well-known, even outside the field of physics.

But your style of writing is just one aspect of how other scholars might get to know you. Most fields are quite large in terms of members of the research community, so in order to carve out your space in the community, you also need to position yourself and your work.

How Do You Position Yourself in the Research Community?

There are of course many approaches you can take to position yourself in relation to a particular research community. But let's try this approach as an exercise. Imagine the research landscape as a playground. In one corner, we have applied research, vibrant, and immediate, where the action takes place – in close collaboration with other stakeholders in our surrounding society. In another corner, there is basic research, quiet and contemplative, a place of thought, and a place where you can focus, question, and dig deeper. The scale and urgency of your research question anchor the other corners, giving this playground four corners. So how do you position yourself? And how do you explore, navigate, and occupy a part of this space? If you move across this playground, you might find yourself drawn to different areas at different times, and that's completely fine.

As we now conclude this chapter, take a moment to envision yourself in this research playground. Where do you see yourself right now? Or 10 years from now? Are you in the midst of the action, testing and iterating, prototyping, and doing interviews and user studies? Or are you more on the sidelines, observing, reflecting, theorizing, and contemplating? Are you trying out solutions? Or are you developing concepts, frameworks, and theories? Are you more interested in emerging and urgent problems? Or are you more drawn to large-scale global concerns? Your position on this playground is not fixed; it can and likely will shift as you move through your academic career. The most important thing is that you reflect on your position, make sure that you focus on the things you find most important, and that you make your time count, contributing to the field of HCI/interaction design in a way that is meaningful to you and beneficial to the community at large.

Given how you position yourself, you can then let this position inform your voice and your academic writing. If you go into urgent problems and applied approaches, then maybe your voice is about offering sets of guidelines, recommendations, and illustrations in the form of cases and prototypes that demonstrate proposed solutions. On the other hand, if your work leans more toward theoretical approaches, then maybe your voice will highlight new concepts and the identification of new paths, approaches, and frameworks.

No matter if you are leaning more toward the urgent and applied, or the basic and long-term, it is of crucial importance for your positioning that you speak with a clear academic voice. Through this clarity, you will add valuable contributions to the research community, and through your individual actions, you will contribute to the collective movement that drives the research community forward.

In the next chapter, we will take a step back from this kind of hands-on chapter that has, to a large extent, been about having a discussion on focus and framing, and the development of your own academic position and voice. Instead, we will look at the fast and slow approach to research from a critical perspective. Accordingly, the next chapter discusses what I call "the temporal dilemma" in interaction design research, and here, we will discuss not only the advantages of fast and slow approaches but also the downsides of each approach.

In short, the next chapter is devoted to a critical perspective where the "fast vs slow" approach is critically examined. If your approach is ultimately about going fast, then how do you know that you are fast enough? Likewise, if the intended process should be slow to allow for in-depth examinations, then how do you know if the pace is slow enough? That is what the next chapter is about.

CHAPTER 7

Critiques – Fast Enough? and Slow Enough?

INTRODUCTION

This book has focused on fast and slow approaches to HCI and interaction design research. It has examined these two temporalities related to applied and basic research, as well as how they relate to the urgency and scale of the research problem. In addition, this book has used this as a temporal lens to discuss ways of aiming and approaching a research topic, and it has covered ways of focusing and framing a research agenda. Still, while this has been good to examine possible approaches and temporalities, little has so far been said about the shortcomings of each approach. This chapter will accordingly seek to outline a critical perspective where I discuss, for instance, if the fast approach is fast enough, and if the slow approach is slow enough. We choose a particular temporality for the research projects we conduct, but what if there is a misalignment between the intended temporality and the problem at hand? In short, what are the potential problems associated with each temporality? Or put differently, how can we critically examine the fast and slow approach? And in doing so, what are the problems or associated shortcomings with each approach?

On an overarching level, this chapter critically examines these temporal modalities within the field of HCI/interaction design research, weighing their pros and cons in various research contexts. It interrogates the assumption that faster is better and challenges the notion that slower

 DOI: 10.1201/9781003343745-7

necessarily means more thorough or thoughtful. The aim here is to illuminate a pathway that harnesses the strengths of both approaches, recognizing when it is beneficial to accelerate the research process and when it is essential to decelerate and go more in-depth.

Ultimately, the goal of this chapter is to encourage a more nuanced understanding of time, duration, and pace in research – a recognition that different research questions and contexts may require different temporal strategies. Accordingly, this chapter advocates for a flexible approach to research pacing, one that adapts to the demands of the topic at hand while remaining mindful of the broader implications and responsibilities of interaction design research.

While Chapters 5 and 6 were kind of hands-on in the discussion on aim, approach, focus, and framing, this chapter shifts the tempo, and also shifts from the straightforward approach, to more discussions and reflections. In doing so, it shifts from stating what needs to be in focus to a more reflective position where I discuss and ask questions, rather than presenting a set of guidelines and things to think about. With these shifts in mind, it is now time to enter this chapter.

Fast Enough? Or Slow Enough? – Paces within Interaction Design Research

The discourse on pace within interaction design research is one of balance – between the swiftness necessary to keep up with rapid technological advancements and the thoroughness required for sustainable and impactful scholarship. At the center of this discussion lies a dichotomy: fast methodologies that prioritize speed and adaptability vs. slow methodologies that value depth and consideration.

In the fast approach, there's an emphasis on keeping pace with the rapid rate of change inherent in technology. Over the last 20 years, we have been through the development of the web, mobile technologies, Web 2.0, IoT – Internet of Things, and generative AI, just to mention a few technology trends. Advocates of this approach argue that quick responses are necessary to remain relevant and to take advantage of fleeting opportunities for digital innovation. This speed allows researchers to explore, test, prototype, and iterate quickly, to fail fast and learn faster, and to bring new designs into reality in shorter cycles. However, this urgency to move at an accelerated pace can sometimes result in surface-level understandings that fail to appreciate the complexity of human–technology interactions. It may for instance overlook the nuanced impacts on society, culture, and individual behavior, favoring immediate utility over long-term value. Further on, it might oversee inequalities, and who are privileged users of these new technologies, and even worse, who is rendered invisible or even used in the transformation toward a new digitalized reality.

On the other hand, the slow approach enables research with deliberate pacing, longitude, depth, continuity, and reflection. This methodical approach allows for comprehensive explorations that consider the long-term implications of design decisions and the use and adaptation of new technologies – over time. Here, research is unhurried by the pressures of rapid innovation cycles, providing the space needed to examine the broader societal implications and ethical considerations of the -design and use of new technologies.

Yet, critics of the slow approach question its feasibility in a fast-paced world. There is a risk that by the time thorough research is completed, the technology or context it addresses has changed or evolved, rendering the insights less applicable or entirely obsolete. Further on, the slow approach is sometimes criticized for being too introverted and not paying enough attention to the rapid shifts in society, and as such, it is sometimes being labeled as "armchair philosophy" – an approach typically described as a way of providing new developments in a field that does not involve primary research or data collection – but instead analysis or synthesis of existing published research.

The Temporal Dilemma in Interaction Design

In the contemporary landscape of interaction design, the tempo at which research and development proceed presents a significant dilemma. On one side of the spectrum, there are "fast enough" methods that cater to the market's relentless demand for rapid innovation. These approaches are often celebrated for their rapid pace and the seeming agility they provide in a fast-moving technological environment. Proponents of such "fast" methods argue for the ability to quickly bring new products and ideas to fruition, thus maintaining a competitive edge and relevance in the eyes of consumers and stakeholders.

Yet, this "fast-paced" race to the future is not without its critics. The haste inherent in "fast enough" methodologies can lead to a superficial engagement with complex social issues. Design solutions emerging from this rapid pace may lack depth, failing to address the underlying structures, situations, and systems that shape user experiences. These solutions risk being ephemeral, possibly becoming obsolete as quickly as they emerged, or failing to achieve sustainable integration into the fabric of daily life due to their shortsighted development process.

In stark contrast, "slow enough" methodologies embrace a different temporal rhythm, treating time as a resource and opportunity rather than a limitation. This approach is characterized by reflection, depth of thought, and thorough investigation into the long-term implications of design choices and how the use of digital technologies changes over time, gets embedded in everyday activities, and becomes part of the structures that govern and produce our everyday lives. Here, the research process unfolds gradually, allowing for comprehensive considerations of how design and the technologies we design interact with and impact society. Proponents of this method argue that by taking the time to deeply understand the contexts in which design operates, we can create solutions that are not only more robust and ethically sound but also have a lasting positive impact. In short, by going slow, we can slowly but deliberately move toward a better future.

The "slow enough" approach suggests that design research processes born out of careful contemplation are more likely to account for the complexities and diversities of human behavior and societal needs. It is posited that these methodologies yield results that are not just reactionary or about fulfilling user requirements but are proactively shaped to adapt to future changes in societal values and norms. In my view, both "fast" and "slow" approaches have their place in the interaction design research ecosystem, and perhaps the optimal path lies not in choosing one over the other but in finding a harmonious balance or combinations that leverage the strengths of both. This balance could involve alternating between the two methodologies at different stages of the design research process or finding innovative ways to combine their merits within a single research project.

Ultimately, the temporal dilemma in interaction design is an ongoing issue for any interaction design researcher. Should you align your research with the latest technology trend, or should you seek a more long-lasting research agenda? For sure, these can also be combined. For instance, and if coming back to an example already introduced in this book, if you are interested in AI, but you want to address it through the lens of ethics as an almost eternal philosophical topic, you can choose to focus on AI ethics, what it is, what it needs to be about, how ethical AI systems should be built, what an ethical perspective on AI implies for design, practice, and society at large, etc. For sure, this is a position you can take where you combine a fast approach to follow and track the most recent developments in AI while staying focused on ethics as a more long-term perspective. Yet, this is also a choice. No matter if you go for the fast approach, the slow approach, or if you decide to combine the two, it is a choice you have to make.

Overall, this dilemma reflects larger conversations about the role of technology in society and the responsibilities of interaction designers. It invites contemplation on how best to align design processes with both the present needs and future aspirations of the people, contexts, and communities they serve. Further on, this discussion can be said to advocate for a conscientious approach to interaction design that considers the ramifications of speed and the virtues of slowness in creating technologies and experiences that enhance, support, and enrich human life.

The "Fast Approach" Reexamined – The Drawbacks of "Fast" Methodologies

The "fast" methodologies in interaction design research, characterized by their rapid response and adaptability, often receive accolades for their efficiency in addressing urgent issues. The ability to quickly produce solutions that respond to immediate technological changes and user demands is undoubtedly valuable in today's fast-paced digital environment. Yet, such methodologies invite scrutiny. In short, if you are "going fast" in your research, then how do you know you are going "fast enough"? If you are continuously chasing the next big thing, how do you know you have the speed to follow a rapidly developing phenomenon?

Further on, critics point out that the haste inherent in these fast approaches can sometimes lead to results that lack depth and fail to consider a comprehensive range of social and environmental factors. The intense focus on speed and quick deployment may overlook the need for solutions to be sustainable and adaptable to more long-term societal shifts. This can lead to design outcomes that might only be relevant or effective in the short-term.

Additionally, the drive to keep up with the latest technology trends can sometimes prioritize novelty over necessity, resulting in solutions that may not stand up to future technological advancements or changes in user behavior. There is a danger that such outcomes become quickly outdated or irrelevant, necessitating further rounds of rapid development that may not address the root causes or systemic issues within the design problem space.

To critically examine fast methodologies thus requires a critical balance – recognizing the value of promptness and adaptability while also considering the importance of thoroughness and foresight. This balance ensures that while immediate design needs are met, the solutions developed are also robust, considerate of wider impacts, and built to last beyond the particular context in which they were created.

Overall, this suggests that although you might be leaning toward a fast approach, you might need this balance, and accordingly, you might need to complement fast-paced activities with other project activities that are more geared toward a slower pace. On the other hand, the slow approaches also have some drawbacks that need to be considered.

The "Slow" Approach Reexamined – The Drawbacks of "Slow" Methodologies

The "slow research" approach in interaction design, which emphasizes thoroughness and deliberate pacing, certainly has its proponents. Typically, so-called "slow scholars" (see e.g., the work by Hartman & Darab, 2012; Mountz et al., 2015; Bozalek, 2017) argue for depth over speed, advocating for a research tempo that allows for comprehensive understanding and integration of insights and solutions into broader social contexts. This slow approach, focused on reflection and methodical progress, aims to produce outcomes with lasting relevance and adaptability.

However, in an age where technological advancements occur at rapid speed, when everything is about change, and when our surrounding society is rapidly transformed, the slow approach faces criticism. Skeptics caution against the potential for research conducted at such a slow pace to fall out of sync with the immediate, real-time needs of society. They argue that the rapid iteration cycles of technology development and deployment may not afford the luxury of a slow research process. Here, "knowledge" is only important to gain if it can also be applied. Accordingly, applicability is the key critique of the slow approach.

There's also concern that the substantial time and resource investment required for slow research could limit the exploration of a broader range of possibilities, inadvertently neglecting emergent technologies or innovations that could be central to the field. Critics worry that an approach that is too slow and leisurely in nature may miss opportunities for significant impact due to its extended timelines.

The critique calls for a careful examination of the slow research paradigm, urging a reassessment of its applicability in a rapidly evolving field.

Here, a discussion is needed on how to balance the depth and quality of research that takes time to mature with the agility needed to stay relevant in a fast-paced technological landscape.

Methodological Adversaries

In the area of interaction design research, methodological choices are always associated with challenges that can significantly impact the efficacy and outcomes of scholarly work.

The fast research methodology, prized for its rapid output, often faces the obstacle of project overload. Researchers working under the pressure of tight deadlines and the demand for quick deliverables may find themselves spread too thin. This situation risks not only the quality of the research outcomes but also the well-being of the researchers themselves, potentially leading to burnout. In short, there is always a new tight deadline, and there is always something new to attend to.

On the flip side, slow research methodologies, while allowing for comprehensive exploration and thoughtful conclusions, come with their own set of challenges. The luxury of time can sometimes transform into a curse of inertia, where decisions are endlessly deliberated upon, and actions are postponed. In a rapidly evolving technological domain, such prolonged contemplation can result in missed opportunities and research findings that may no longer be relevant by the time they are published.

The challenge of finding the right tempo and balance between these methodological approaches is further compounded by the diversity of stakeholder interests involved in research projects. Stakeholders, whether they are closely related to the research project or more remotely associated, often come with their own expectations and pressures, which may skew the research process toward one end of the methodological spectrum or the other. Researchers must deal with these competing interests, advocating for the methodological rigor that best serves the research objectives while also addressing the practical needs and timelines of these stakeholders.

Arriving at a balanced approach, and planning of the research process that honors the depth and integrity of slow research while maintaining the relevance and applicability of fast research, requires a nuanced understanding of the research context and a flexible approach to methodology. It demands that the researcher has an adaptive strategy that considers the unique demands of each project and the particular needs of all stakeholders involved. Balancing these methodological adversaries is a delicate task but one that is essential for producing robust, timely, and impactful research in the field of interaction design. For sure, design is always carried out on behalf of someone and for someone, and accordingly, this means that design research always has a set of stakeholders involved, no matter if it is explicitly stated or not.

In the next section, I discuss the fast vs. slow approach from the critical perspective of how well it supports innovations, reflections, and analysis.

Between Rapidity and Reflection

In interaction design research, the pace at which we move forward can significantly influence the quality and applicability of our work. As illustrated in this book, the choice between adopting "fast" or "slow" research methodologies is nuanced and cannot be reduced to a simple binary choice. Instead, it's about finding a balance – a thoughtful combination that honors the strengths and addresses the limitations of each approach. Even if you have a 10-year research program, you can choose to have 2–3 overarching research goals which you can return to over a decade, while still being able to organize the research program as smaller and faster "modules" where each short module might address something urgent or something that is rapidly evolving.

As I have described in this chapter, the rapid pace of "fast" methodologies is crucial when the goal is to stay abreast of the latest technological advancements and to meet the immediate needs of the market. The ability to quickly iterate, prototype, and test can lead to innovations that are timely and directly applicable, keeping researchers and practitioners in sync with the pace of change. Conversely, "slow" methodologies provide the space for deeper reflection and critical analysis, allowing for the thorough exploration of ideas and consequences. This slower pace can ensure that research findings are not just innovative but also ethically sound, socially responsible, and sustainable over the long term.

However, the dichotomy between fast and slow approaches to research is more of a spectrum than a split. Although I have in this book kept them apart for analytical reasons, it should be noted here that many successful research projects are successful because they have found a way of integrating fast and slow elements into the overarching research design. It might be the combination of many small, rapid projects with more long-term efforts to theorize, or it might be through iterative processes where a core question is revisited over time, through many shorter iterations. Here, an adaptive approach is crucial to find a balance and an integration that works for the particular project at hand.

Further on, this adaptive research methodology recognizes that different projects may require different approaches at different times. Flexibility is key. A project might start with rapid exploratory studies to gauge the potential of an idea and then transition into slower, more contemplative phases where the implications of that idea are fully explored.

This adaptive approach is particularly effective in interaction design research, where researchers must often balance the need for quick responses with the complexity of human-centered (or even more-than-human) design problems. By constantly being selective and aware of decisions that imply a trade-off between rapidity and reflection, thoughtful design researchers can ensure that their work remains both relevant to current technological trends, while being deeply informed by human and non-human needs and behaviors.

In seeking this balance and integration, HCI and interaction design researchers should be strategic in their planning, continuously assessing the demands of the research question at hand. They should be prepared to shift gears, speeding up when surrounding society or technological pressures call for it, and slowing down when the depth of inquiry demands more time.

Ultimately, striking a balance between the fast and the slow approach to interaction design research demands sensibilities and thoughtfulness. It requires a nuanced understanding of the research context, a clear sense of the project's objectives, and an ability to adapt to emerging insights and circumstances. It's about moving forward with intentionality, whether that means making swift progress or taking the time to ponder and probe the deeper questions at the center of our socio-technological future.

Discussion – The Temporal Nature of Knowledge in Interaction Design Research

So, what does all of this imply for research as an act of generating new knowledge? Can knowledge be outdated? Is there something like timely knowledge? Is knowledge time-critical? In short, what is the relationship between this integration and combination of fast and slow research approaches on the one hand, and knowledge on the other hand?

In the ongoing technological and societal transformation, the temporality of knowledge emerges as a critical concern. The question of whether knowledge is time-dependent is related to the core upon which interaction design research is built. Is there, in the expansive and ever-evolving field of design, something we can define as "timely knowledge," or is the very notion a mirage, transient, and elusive? Here, I would suggest that the concept of "timely knowledge" implies an intrinsic value placed on information that is most relevant to the current zeitgeist. It suggests that the utility of knowledge has a shelf life that diminishes unless it continues to resonate with or reflect ongoing changes. In the modern fast-paced world where design interventions can quickly become obsolete, the pressure to constantly update and refresh one's knowledge base is essential. Accordingly, I would like to suggest that in interaction design research, this pursuit of contemporaneity is not just a mere academic exercise; it is a vital ingredient for relevance, innovation, and impact.

However, the assertion that knowledge can become outdated seems to imply a linear progression of information and knowledge, where new insights consistently build upon or outmode old ones. This viewpoint might however be overly simplistic, failing to recognize the cyclical, contradictory, and sometimes recursive nature of knowledge. In many instances, what we perceive as "new" knowledge often stands on the shoulders of the "old." The traditional dichotomy between the "obsolete" (the old) and the "cutting-edge" (the new) knowledge accordingly fails to account for the depth and breadth of understanding that come from integrating historical insight, paradoxes, dilemmas, contradictions, and forward-thinking.

Consider the concept of "timeless principles" in design – ideas that possess enduring relevance regardless of technological advances. Principles of usability, for instance, can transcend the particulars of the medium or the moment. The human factors, including ergonomics, perception, cognitive abilities, eye–hand coordination, etc., that drive effective UX design do not rapidly depreciate over time but rather evolve, combining with new insights to form a richer understanding.

It could be argued here that knowledge is in a state of perpetual evolution, not necessarily invalidated by age but rather augmented and refined. Thus, the obsolescence of knowledge is not a default condition but a consequence of its failure to adapt and interweave with new discoveries and societal shifts. The idea that knowledge is replaced suggests a paradigmatic shift, something similar to a Kuhnian revolution, where frameworks and understandings are not just extended, but fundamentally transformed. In such cases, the "old" knowledge is not so much outdated as it is subsumed into a greater, more comprehensive understanding.

Within the area of interaction design research, this ongoing conversation between "old" and "new" knowledge frames a dynamic temporality where researchers must cultivate a discerning eye for what knowledge to preserve, what to modify, and what to set aside. It is through this discernment that knowledge proves its worth – not solely through its novelty but through its enduring applicability and its capacity to be repurposed and recontextualized in light of new studies, insights, and changing conditions. For sure, the most recent turn in HCI and interaction design research toward "more-than-human" design might be one such transformation where interaction design researchers are now questioning and rethinking things like "who is the user?", "who are we designing for?", "for whom is the design working?", and "for whom is it sustainable?" In changing perspectives, we also redirect our knowledge interests, and accordingly our research efforts.

In grappling with the notions of "timely knowledge" and "timely contributions," we must also consider the mechanisms by which knowledge is disseminated and the speed with which it permeates practice. The rapidity of this transfer – and the industry's appetite for instant application – can sometimes outpace the slower, more deliberate pace of reflection, and critique that ensures depth and rigor. Still, it is not only our responsibility as researchers to produce new knowledge and research results for further dissemination to industry; it is also our responsibility to make intellectual turns and think, rethink, and crave out new research directions. These turns take time.

If we now leave this overarching discussion on timely knowledge and reflections on ways of bridging and integrating fast and slow research approaches, we can now revisit the "Double Diamond" design research model from this critical perspective. In the next section, we accordingly examine this model of the design process not only from the perspective of how it defines the process as four distinct phases, but also how it, in doing so, fails to address how the pace and temporality actually unfold in many design research projects. For instance, design researchers very seldom work along a linear process. Instead, they typically go back, iterate, explore, think, and rethink in order to move forward. This, and several additional factors, is covered in the next section.

Revisiting the Double Diamond Model – A Critical Perspective

As described in Chapter 4, the Double Diamond model is a visual representation of the design process, first proposed by the British Design

Council in 2005. It illustrates a design methodology that is split into four main stages – Discover, Define, Develop, and Deliver – across two "diamonds." The first diamond represents the problem space, and the second represents the solution space.

Although the Double Diamond model has been widely adopted as a framework in design processes, it is not without significant critiques. The model represents a structured approach to problem-solving through two phases of divergent and convergent thinking. However, several issues have been raised by designers, researchers, and practitioners regarding its limitations in capturing the complex, dynamic nature of real-world design.

Its oversimplification of the design process is among the most common critiques. *Discover, Define, Develop*, and *Deliver* are the four discrete phases that make up the Double Diamond's orderly evolution of design. But in practice, the design process rarely proceeds in such a straight-line and predictable fashion. The actual design process is frequently messier, requiring feedback loops, backtracking, and numerous iterations. The various phases may also take longer or shorter times, as I explained in Chapter 4. The model's inability to take this fluidity into consideration may cause people to view design as a process that proceeds in a clear step-by-step fashion while neglecting the frequent return to previous phases. Critics contend that designers, particularly those with less expertise, may be misled by this straightforward portrayal into believing that once a phase is complete, there is no need to revisit it – a dangerous assumption in the iterative nature of good design thinking.

Additionally, the Double Diamond underemphasizes *iteration*, a key idea in design. Divergent and convergent thinking are undoubtedly acknowledged by the model, but it makes no mention of how crucial it is to keep going back and improving concepts as you go. Before reaching a final solution, successful design frequently necessitates several cycles of brainstorming, prototyping, and feedback. Design results must be improved and refined through this iterative process, which incorporates feedback consistently. A designer's capacity to efficiently refine solutions may be hampered by the model's failure to make this apparent, which can lead to a restricted grasp of how iteration works both within and between stages.

The Double Diamond's ambiguity in its stages is another problem. The transition from Define to Develop and Discover to Define are not always straightforward. It might be challenging to tell when one phase finishes and another begins because the stages sometimes overlap or become blurry. For example, it may take more than one discrete period to go from comprehending the issue space to precisely characterizing it. Likewise, it might not always be easy to move from development to delivery, which begs the question of how designers can distinguish between these stages. Teams trying to adhere strictly to the model may encounter difficulties as a result of this ambiguity, particularly when dealing with intricate, multi-dimensional issues that do not cleanly fall into the phases specified by the framework.

The methodology has also drawn criticism for not being user-centric at any point in the process. The Double Diamond is less clear about sustaining this engagement across the following phases, even while it encourages designers to interact closely with users during the Discover phase to comprehend their demands and motivations. Maintaining the design's alignment with user needs and expectations across the Define, Develop, and Deliver stages requires constant user input. The Double Diamond might unintentionally encourage a more solitary approach to decision-making following the completion of the initial study by undervaluing this continuous interaction, which could result in a misalignment between the solution and user requirements.

The possibility of confirmation bias is an additional worry. The Double Diamond's structure may unintentionally cause designers to concentrate on information that supports the presumptions they made in the Discover stage. This danger occurs when designers, whether intentionally or inadvertently, give weight to input that confirms initial theories and ignore conflicting data later on. Because designers may become obsessed with solutions that support their original discoveries rather than keeping an open mind to fresh insights that surface later, these biases can restrict the scope of innovation and creativity. This problem emphasizes how crucial it is to cultivate an attitude that welcomes contradicting data in order to improve and refine design results.

Because the Double Diamond is based on a Western-centric view of design, it also has cultural and contextual restrictions. Its use may not be generally suitable or flexible enough to fit into different organizational or cultural situations. The Double Diamond's design philosophy tends to mirror the norms and values of business settings in the West. These may not be in line with the procedures or requirements of companies in culturally diverse contexts, where approaches to problem-solving might be more collaborative, iterative, or shaped by alternative philosophical frameworks. Models that consider various cultural viewpoints and approaches are becoming more and more necessary as the design industry becomes more worldwide.

Furthermore, one major flaw in the Double Diamond model is its disregard for post-delivery factors. The paradigm provides minimal direction for handling the crucial post-delivery stage, where ongoing user input, assessment, and improvement are required for sustained success. The notion that the design process concludes with delivery ignores the significance of continuous user involvement, particularly in a world where products must change quickly to meet evolving user needs. If this important stage of the model is skipped, designs may not be able to adapt to changing user needs after they are first introduced.

Lastly, as practical issues, the Double Diamond model's resource intensity and lack of integration with project management methodologies have been brought to light. It can consume a lot of resources, especially during the Discover phase, which calls for in-depth investigation, user interaction, and exploration. For some projects, especially those with limited funds or schedules, this might not be possible. Furthermore, some project management techniques that are frequently employed in business settings are difficult for the Double Diamond to interact with. When attempting to match design processes with more general business and project management frameworks – where responsiveness and agility are crucial – this may cause problems.

In summary, even though the Double Diamond model has gained popularity due to its structure and clarity, it is important to critically reflect on its drawbacks. When using this model in practice, it is crucial to take into account a number of problems, including its cultural biases, lack of iteration, stage ambiguity, oversimplification, and underemphasis on user participation. A more sophisticated and adaptable approach to design is promoted by critically analyzing these constraints, guaranteeing that the procedure can adjust to the intricacies and unpredictabilities present in actual design problems. However, despite these criticisms, the Double Diamond model remains a valuable tool in the design industry. It provides a clear framework for understanding and applying the design process, even if it needs to be adapted for different projects and teams. It's important to use the model as a guide rather than a strict rulebook, adjusting the process as needed to fit the unique needs of each design research project.

CONCLUSION

This chapter has served as a critical examination of the pace of research within interaction design, and in doing so it becomes clear that the constructed dichotomy between "fast" and "slow" methodologies might be a false one. The field of interaction design, inherently interdisciplinary and dynamic, requires a spectrum of integrated approaches, finely tuned to the unique challenges and opportunities of each design research project. In rejecting any binary view of fast vs. slow approaches, this chapter advocates for a balanced, relational, and accordingly integrated methodology that adapts nimbly to multiple goals, multiple research temporalities, multiple stakeholders, technological advancement, user needs, and societal changes.

The discussion in this chapter has, on the one hand, been in the form of a critical examination of the fast vs. slow research approach, but on the other hand, it has, on a higher level, been about a foundational shift toward a deeper understanding of temporal dynamics in interaction design research. In rejecting the binary view where researchers have to make a radical choice between a fast and a slow approach – to be applied, or go for a "slow scholarship" – to instead examine what a fruitful integration of the two temporalities could offer for the particular research project at hand, the discussion also moves from an instrumental choice to more practical research planning. In acknowledging this, it becomes clear that this balance is not static but rather a fluid process that oscillates to the rhythm of research questions, stakeholder needs, and project goals.

This nuanced approach to temporal strategy in research planning also aligns well with the core values of interaction design – values that place human experience at the forefront of technological progression. Inclusivity, sustainability, and ethical integrity are not mere buzzwords but guiding principles for interaction design research. These principles demand methodologies that are iterative and responsive and approaches that allow for design explorations that are resonant with users and responsible toward the broader ecosystem, including both human and more-than-human actors.

In approaching the end of this chapter, we can now take a step back and reflect on the position that knowledge is neither inherently time-dependent nor immune to obsolescence. In my view, it is tied to the researcher's ability to continuously contextualize, critique, and reframe it within the evolving narrative of human and more-than-human technology interactions. In this regard, interaction design researchers are not just creators of knowledge but also knowledge curators, tasked with the responsibility to let it grow, but also change to acknowledge or drive change in society. Here, knowledge itself might be the ultimate design material – we learn and we use this learning to move forward over time.

Thus, we return to the field of interaction design research with a renewed understanding – knowledge, like the designs it informs, is not a static entity but a dynamic interplay that is as timeless as it is timely, positioned between history and the future. Accordingly, and in line with these concluding reflections, the next chapter is oriented toward another aspect of fast and slow, and that is the question concerning whether the timeline is already set or not? In short, if we can "make time" for our research, and if major breakthroughs in research also enable or imply temporal jumps into a future that was estimated to be much further away. To "make time" or "break time" through breakthrough research is strongly dependent on strong and clear research *visions* and *prospects*. Accordingly, the next chapter is devoted to these central concepts.

CHAPTER 8

Making Time, Breaking Grounds

INTRODUCTION

So far, this book has addressed fast vs. slow approaches to design research projects from the perspective that the flow of time and approaching deadlines are non-negotiable. Time passes, and the only thing we can do is choose between fast or slow approaches and maybe sometimes seek to combine the two to ensure that our research and its contributions are timely.

But what if we can break or reconfigure this flow of time? What if we can do temporal leaps, get more time, or even be ahead of time? In fact, is time really this linear, evenly distributed, and non-negotiable constant? Or are there ways of making and breaking time? This chapter, called "making/breaking time," sets out to explore these questions, with a particular focus on design research and the importance of *visions* and *prospects* in design research. For sure, in design-oriented research, change and being future-oriented is at the very core of the approach. Here, I suggest that strong visions, together with a design approach, can enable us to be ahead of time. An appropriate design approach can 1) intellectually put us in a future situation where we can think about how that would be, and 2) through deliberate actions taken, we can then not only envision but also contribute to the making of these alternative futures.

Further on, well-developed prospects enable us to develop our thinking on how to get there – on how to reach or realize these visions for the

 DOI: 10.1201/9781003343745-8

future. By managing to do this, to formulate strong visions and prospects, there is an alternative to timely research beyond "deliver on time." Here, *timely research* might even provide more time for us. In fact, powerful design research can enable us to get to the future, or at least see glimpses of what an alternative future could look like. This moves the discussion from a narrow focus on timely research as a matter of doing it "while there still is time" to also consider what timeliness is fundamentally about, and whether there is such a thing as timeless research.

The previous chapter, that is Chapter 7, was about applying a critical perspective on fast and slow approaches. Now, this chapter is about applying one such critical perspective on time and temporality. In doing so, I suggest that maybe time, timeliness, and deadlines are not something purely given by nature but maybe can be thought of as also being socially constructed. Sometimes we feel lots of stress, think we are too late, and think that we are in a hurry, when in fact there is plenty of time. Still, no matter if we see time as something given, something that we cannot influence, or if we see time as socially constructed, it is still important to "be on time" if seeking to do timely research. Here, this "being on time" also adds a social and collective level to this discussion. Whether or not someone is considered to "be on time" is related to that larger group of people who might have expectations for when you will deliver your results, make a contribution, or add to an ongoing conversation in a field of research.

In coming back to this topic of "timeliness" in research, I want to propose that in the area of design research, "timeliness" extends far beyond the mere synchronization with current trends. In this chapter titled "Making/Breaking Time – on Visions and Prospects in Design Research," we look at the intricate relationship between time and research within the context of design research. I hope this chapter invites you as a reader to engage with questions concerning how interaction design researchers not only align with the emergent cultural *zeitgeist* but also proactively participate in sculpting the future of design.

Further on, rather than perceiving the future as a distant, fixed destination, this chapter redefines time as something that is constantly changing and constantly in the making. Research, in this perspective, is not a race toward a deadline or an inevitable future but an active process of creation. Through the actions we take as researchers, we also form the future of our research area, and how we configure our research process also determines how much time we have. Accordingly, researchers in the field of HCI/interaction design can wield their studies, prototypes, and innovations as instruments to carve out new possibilities, effectively making and breaking time. One thing is to think about the future, but it might be more powerful to seek to invent it – through prototyping alternative futures, through new methodological approaches developed, and through the development of new theoretical lenses that enable us to see things differently. For the design researcher, the research process is thus a *constant negotiation with time*. It involves understanding the currents of cultural and technological change, identifying the opportune moments to introduce new concepts and new technologies, and enabling new socio-technical practices. This chapter dissects strategies for syncing research with these currents, emphasizing the importance of adaptability and future orientation.

Furthermore, the discussion in this chapter on ways of making and breaking time challenges the notion that research should always be reactive and deadline-oriented – as if the timeline was already given. In doing so, this chapter advocates for a proactive stance where researchers craft visions and prospects – thoughtful anticipations of preferred and desirable futures that inform the direction and purpose of their work. By focusing on visions and prospects, the design researcher can contribute not just to the immediate landscape of design but also help lay the foundations for its evolution – over time – beyond just contributing with descriptions of "what is," and instead also be normative about what could, should, or what ought to be.

In this context, it is not only about the timing and duration of a project, but it is ultimately about making timely contributions. But how? Well, there are two options here. You can either try to "catch the wave," to identify an important research problem, and contribute to this wave while it is still interesting and relevant. Or you can formulate a research vision and prospect and then seek to contribute to that vision. For the first approach, it is about contributing while there is still time. For the second approach, it is also about making time for the research that needs to be conducted. If the vision is grand, it might imply that lots of time needs to be spent to get there.

Accordingly, for a strong research prospect to work, it needs to be 1) future-oriented, in that it should describe a future preferable situation, and it needs to be 2) situated in the future, so that there is time between now and this future state, a period that provides a temporal space for some research to be conducted on this topic (the "making time" aspect). In following this approach to "futuring" (for approaches to futuring and future inquiry see for instance Cornish (2004), Sturdee and Lindley (2019), Bendor et al. (2021), Oomen et al. (2022), Søndergaard et al. (2023), and Cooper (2024)), we need to shift from traditional temporal markers to a conceptualization of time as a dynamic continuum. Here, researchers have the agency to influence and shape the path of design, ensuring that their work is not only responsive to the here and now, but also resonant with a preferable future. Accordingly, this shift toward not only describing, but more actively contributing to the forming of our future, also comes with larger responsibilities and a call to collaborate with others in this effort. Ultimately, the discussion culminates in a call for a broader dialogue – a dialogue that includes not just researchers but practitioners, stakeholders, and the wider community. This conversation is about bridging the present with the future, and it is about making strategic choices that are grounded in deep understanding while reaching toward the expansive possibilities that lie ahead in the field of design (for related work on this ambition, see for instance Wiberg, 2014a; Kozubaev et al., 2020).

Again, coming back to this notion of "timeliness," I suggest that in the quest for impactful research within the field of design, the concept of "timeliness" assumes a role far more complex than a simple alignment with the latest trends. Accordingly, we need to critically dissect the varied strategies by which scholars in design-oriented research can both attune their efforts to resonate with the emergent cultural *zeitgeist* and, with greater ambition, shape the trajectory of the future. Along these lines, this chapter challenges the traditional view of the future as a predetermined endpoint that researchers inevitably converge upon. Instead, it presents the future as a dynamic and changeable landscape, formed by the researchers whose inquiries and innovations become the tools for informing and establishing what is yet to come.

The concept of *zeitgeist* (see e.g., Krause, 2019; Fruehwald, 2017; Boring, 1955), central to ways of doing timely design, encompasses the prevailing cultural, intellectual, ethical, and political climate of an era. It is about a collective consciousness that influences and is influenced by the creative expressions within design. In design research, the zeitgeist is both a guiding force and a reflection pool. It demands sensitivity to current and contemporary societal narratives while also providing a context for the work produced. A design that *resonates with the zeitgeist* is one that speaks to the immediate preoccupations and aspirations of the people it's meant to serve. It taps into the shared feelings and ideas that are omnipresent yet often unarticulated.

Incorporating the notion of zeitgeist in the formulation of research visions and prospects requires researchers to immerse themselves in the currents of contemporary culture – in the time of now. It involves engaging with the arts, politics, societal debates, and technological advancements that are shaping our social and material world. This engagement ensures that design is not created in a vacuum but is deeply rooted in the lived experiences of the time. Furthermore, for design research to remain relevant and impactful, it must adapt to the evolving zeitgeist, balancing the permanence of human-centric principles with the dynamics of societal change. As such, the zeitgeist is not just a backdrop for design research; it is a lens through which the relevance and efficacy of research are evaluated. Recognizing and interpreting the zeitgeist is a skill that allows designers to create work that is both of the moment and has enduring value beyond the current moment in time.

If becoming aware of the zeitgeist, you as a design researcher can see how you can also deviate, to challenge the existing, and to ultimately make/break time by establishing new paths of research, formulating new research visions, goals, and aims, and by challenging and rethinking "truths" and concepts that we have developed, established, and taken for granted in the zeitgeist.

Synchronizing Research with the Zeitgeist

A strategy here is to synchronize your research efforts with the current zeitgeist. Here, the notion of aligning your research with the *zeitgeist – the spirit of the time* – is enticing for its promise of immediate relevance and recognition. Researchers who adeptly catch this wave of collective interests, emerging trends, and overarching movements in a field of research can indeed make significant contributions that resonate with the contemporary audience. Such synchronization requires not only an acute awareness of current trends but also the agility to respond quickly and effectively to the unfolding needs and interests of society. If managed well, it is an effective strategy to also speed up a research career. You listen in, and you contribute – to the current moment in time.

Synchronizing your research with the zeitgeist involves a sensibility for what is important at the times – a pulse-taking of societal moods, trends, and emergent patterns in society, culture, and technology advancements. For researchers in the field of HCI/interaction design, tapping into this collective movement means their work not only engages with the current discourse but also speaks directly to the shared experiences and concerns of the contemporary audience.

Achieving this synchronization, however, is however no small effort. It requires an attuned sensitivity to the cultural, social, and political currents that shape the lived reality of the population. Any researcher who seeks this alignment must immerse themselves in the present, maintaining a vigilance that allows for the anticipation of emergent trends, needs, and interests. This vigilant state, however, is not just about observing trends but about understanding the underlying forces that drive these trends. In addition to this, responsive agility is crucial. As the zeitgeist shifts with societal changes, researchers must adapt with speed and precision. This agility enables design research to be reflexive, providing solutions, and insights that are not just timely but also forward-thinking, anticipating the trajectory of these societal shifts.

Synchronizing with the zeitgeist is also more than mere alignment; it is an active engagement. It requires the design researcher to be culturally literate, informed, and contextually savvy. This synchronization ensures that research outcomes do not simply mirror the current state of affairs but critically engage with it, contributing to a future that is being shaped by the very research that seeks to understand it.

Forming the Future through Visions and Prospects in Design Research

In the pursuit of defining the future through the lens of design, researchers engage in an act far more creative than mere prediction – they engage in the active craft of shaping and proposing it. The formulation of research *visions* and *prospects* accordingly represents a deliberate approach to construct a desirable future, one that reflects a high degree of aspiration for design principles and embodies our core human values.

This creative act of formulating research visions and prospects extends beyond envisioning what is likely to occur based on current trends. It is about constructing a narrative that sets a clear direction for research efforts – a narrative that infuses the collective research direction with purpose and intention. When researchers develop and establish visions and prospects, they create what can metaphorically be described as a guiding light that illuminates the paths toward envisioned futures. To articulate and actively form the future, design researchers must thus move beyond the role of passive observers and become active creators and stakeholders. They must harness their knowledge, their insights, and their creativity to imagine a world that does not yet exist but one that can be brought into being through the diligent and thoughtful application of design research.

Given this importance of visions and prospects in the research process, we will now look more closely at the role of such prospects in the design research process.

THE ROLE OF PROSPECTS IN DESIGN RESEARCH

In the design research process, prospects serve some critical functions. First of all, a well-articulated prospect can help outline the objectives and anticipated outcomes of the research process. Second, the prospect can work as a fundamental element in guiding research activities, emphasizing its importance in setting the agenda for inquiry and in shaping the potential applications, explorations, and implications of research results.

Prospects in the context of research processes are detailed descriptions of anticipated outcomes that scholars aim to achieve (Newell & Card, 1985). They are based on a clear understanding of what future goals and scenarios the research is intended to influence or create (Bibri, 2018). However, these are not just idle predictions or wishes; they are actionable objectives that are based on systematic and methodical planning (Alexander et al., 1998). In this context, a critical aspect of successful research planning is allowing adequate time for exploration and analysis. Here, prospects enable researchers to define a timeline that accommodates a thorough study, incorporating periods for data collection, analysis, and reflection (see e.g., Gough & Madill, 2012). This approach ensures that research is not rushed and that outcomes are robust and well-substantiated. Here, the relation between a well-articulated prospect and the temporality of a research project is clear.

Further on, and along these temporal aspects of research processes, the use of prospects moves the researcher's role beyond simply reacting to immediate issues. Instead, design researchers take on the role of creators, actively working toward specific outcomes they have defined. This shift in perspective empowers design-oriented researchers to initiate projects with a clear sense of direction and to conduct research that is proactive, purpose-driven, and future-oriented.

To formulate such prospects also has implications for the formulation of the overarching research questions. Here, future-oriented research questions are vital to guiding inquiry along these defined prospects. Such questions should be designed to lead to outcomes that will be relevant and useful in the future state that the research is targeting and envisioning.

They need to be specific, actionable, and directly linked to the overall objectives the research is intended to accomplish.

In addition to this link between prospects and research questions, we should also acknowledge that methodologies used in these research processes must be aligned with the stated prospects, allowing for an approach that is both rigorous and relevant to the research aim and goals. The chosen methods should be capable of addressing the complexity of the research topic and should be designed to produce results that will be applicable in the future context outlined by the prospects. Prospects not only inform the objectives of research but also ensure that the approach and methods used are appropriate for the anticipated outcomes.

The practice of envisioning design research and articulating such prospects is not only a creative effort but also a strategic one. It requires a comprehensive understanding of both the current state of the world and the potential trajectories of technological and societal evolution. Accordingly, it demands a deep engagement with the values and needs of people, ensuring that the futures we shape are inclusive, sustainable, and enriching for all. In current visionary work in HCI and interaction design research, we see this articulated in more critical perspectives – a deeper interest in, for instance gender, disabilities (e.g., Rosner et al., 2021; Arvola et al. (2023), race (Harrington et al., 2021), Indigenous people (Nordin et al. (2023), and so on, as well as in the current developments of more-than-human perspectives and values (Yoo et al., 2023) where it is not only "the values and needs of people" that matter but also what futures we envision for more-than-human worlds.

By working along this approach, design researchers can contribute to a future that aligns with collective hopes and dreams, toward a world that reflects and further develops our shared values and aspirations. So what is this notion of "future"? And how does it align with the focus of this chapter on the making and breaking of time?

Working with Prospects and Being Future-Oriented

The notion of the future as something we actively form suggests that our current research efforts are akin to laying the groundwork for what is to come. We, as design researchers, are not merely forecasting or adapting to a predetermined future; we are actively forming and informing it, and accordingly, we are influencing its direction with our creative and scholarly contributions. In this book, I suggest that the design research of today, when approached with this mindset, becomes a powerful act of future making. Every hypothesis tested, every model constructed, and every user study conducted becomes part of a grander narrative that we write for the future. It is within the confluence of our insights and creative endeavors that the path forward is carved. Accordingly, it comes as no surprise that our field is now not only future-oriented, but also how "futuring" (Fry, 2009; Cornish 2004; Oomen et al., 2022), design futuring (Kozubaev et al., 2020), and collaborative futuring (Hyysalo et al., 2014) have rapidly established themselves as research approaches, together with approaches for imagining and formulating scenarios (Carrol, 1999; Carroll (2003), design fiction (Blythe, 2014), speculations (Soden et al., 2021), speculative designs (Wong & Khovanskaya (2018; Zhu et al., 2024a), fabulations and design futuring Søndergaard et al. (2023; Nijs et al., 2020), and alternative futures (e.g., Nardi, 2015; Jenkins et al., 2024; Linehan et al., 2014; Forlano & Halpern, 2023).

The image of the future as a dynamic landscape that we shape implies that our role is not passive. Instead, we are actively contributing to the growth of a vast and diverse ecosystem of knowledge, technology, and practice. Our design innovations serve as blueprints, laying the foundation for the structures of the future – structures that may enable new ways of living, interacting, understanding, and relating to the world around us. Adopting this proactive stance requires a deep understanding of the implications of our work, a commitment to ethical practice, and a vision that aligns with the broader aspirations of our society. As we engage with the future as a dynamic landscape, we embrace the responsibility that comes with being designers of preferred futures, ensuring that the changes we drive are beneficial, sustainable, and ethical.

To summarize, one can say the following about visions and prospects in design research:

Visions – Can help us to formulate goals for our research and answer the question of why we are doing the research we do in the first place. In the process of formulating research visions, scenarios can be a useful method to describe, analyze, and elaborate on alternative futures.

Prospects – Enable us to break down the vision into manageable parts and examine what projects are needed to realize the prospects. Once we understand the overarching prospect, we can turn this into research programs – long-term research programs that can connect smaller projects to systematically move forward.

For sure, setting visions for the future and working toward alternative futures through prospects is a big responsibility, and this responsibility calls for strategic processes.

The Strategic Processes of Design-Oriented Research

The focus on timely contributions in design-oriented research necessitates a sophisticated understanding of the present moment combined with a visionary outlook for the future. In short, to have a good sense of how *time* also bridges between the present and the future, including an understanding of how much time it might take to "get to the future" in terms of "changing an existing situation into a preferred future situation" in the words by Prof. Herbert Simon (2019). Accordingly, design researchers must engage in strategic processes that help navigate between immediate relevance and long-term impact. How to do this requires not only a design-orientation, but also strategic processes. These processes begin with an awareness of emerging trends, with a particular eye on societal shifts, technological innovations, and cultural patterns. In these processes, design researchers must be adept in methods that allow them to quickly identify and understand these trends, and position their work to not just react to the *zeitgeist* but to contribute meaningfully to its progression.

In parallel, design-oriented research requires the formulation of long-term visions, anchored not just in the extrapolation of current trends, but also in the identification of enduring human needs and values. Crafting these long-term visions is accordingly about more than predicting; it's about steering the trajectory of research and design toward a future that is more value-based and might for instance, take into account perspectives on inclusion, diversity, and sustainability.

The tools and methods employed in this approach range from foresight workshops that challenge current assumptions to scenario planning that explores a variety of future possibilities. Here, design researchers can utilize design thinking methodologies to ideate and prototype possible futures and engage in speculative design to provoke discussion and debate about the kind of future society wishes to create. For some related work on speculative design approaches, see e.g., Auger (2013), Mitrović et al. (2021), Wong and Khovanskaya (2018), and Barendregt and Vaage (2021).

In coming back to how this requires not only a design-orientation, but also strategic processes, I suggest that design researchers need to position themselves as central responsible actors, in a position to harness the power of design to influence the course of human-technology relations, as well as to be responsible for the side effects of their designs. So, given this strategic and responsible mindset, we can now turn to a set of strategies for making timely contributions through design research.

STRATEGIES FOR MAKING TIMELY CONTRIBUTIONS IN DESIGN-ORIENTED RESEARCH

Within the area of design research, time is often, as I have outlined in this book, the silent structure that dictates the rhythm of scholarly work. Here, researchers are pulled between contributing "in time" – reacting swiftly to the urgent needs of the now – and the decision to "make time" for more contemplative research that might contribute to the crafting of future possibilities enabled through digital systems. This section is dedicated to elucidating the strategies that enable researchers to navigate this temporal divide effectively.

"Time-to-Market Research" – On Contributing "In Time"

To contribute "in time" is to respond with agility to the prevailing currents of needs and innovation. This fast-paced mode of research is characterized by its responsiveness to the emergent needs that the *zeitgeist* uncovers. It calls for a high level of adaptability, a keen sense of the market and societal pulse, and the capacity to generate quick turnarounds in research outputs. Sometimes, this can be thought of as a "time-to-market strategy" when considering how to make timely contributions.

On a general level, the strategy for contributing "in time" involves staying abreast of technological advancements, policy changes, and social transformations. It requires the cultivation of a network of collaborators and stakeholders who can offer real-time insights and feedback, enabling researchers to produce solutions that are both relevant and urgent. Such contributions often manifest as interventions in ongoing debates, technological breakthroughs, the development of new solutions, or responses to emergent societal challenges.

Changing Perspectives – "Making Time" for Thoughtful Exploration

In contrast to the first strategy, which is well-aligned with the immediacy of reactionary research, making time for exploration is an alternative strategy that invites a slower, more deliberate pace. It's about carefully and thoughtfully investing in the future – envisioning what could be and methodically working out a course to get there. This strategy demands a visionary outlook, one that looks beyond the present, seeking to delineate and realize preferable future situations. "Making time" for this type of research can be seen as a purposeful pause from the rush, a deep breath that allows for the development of profound, transformative knowledge. It means embracing uncertainty, diving into the depths of complex problems, and emerging with frameworks, theories, and paradigms that could shape the contours of future societies.

Working across the Temporal Dichotomy between "In Time" and "Making Time"

The key to working across this seeming dichotomy between "in time" and "making time" lies in the ability to balance the demands of "fast" and "slow" research. To be successful here, the researcher must acknowledge the importance of timing, recognizing when to *accelerate* to meet an immediate need and when to *decelerate* to ensure depth and foresight in their work. This balancing act is not static but a fully dynamic process.

One strategy here is the use of *Forecasting and Trend Analysis*. Researchers can put themselves in a position to foresee changes in their field in the future by utilizing tools like trend analysis, environmental scanning, and Delphi investigations. They can modify their research pathways proactively rather than reactively thanks to this forward-looking strategy. For example, environmental scanning assists in identifying external elements that may have an impact on the research, while the Delphi technique uses expert opinions to foresee possible outcomes and trends.

By leveraging tools such as Delphi studies, environmental scanning, and trend analysis, researchers can position themselves to anticipate future shifts within their field. This forward-looking approach enables them to adjust their research trajectories proactively, rather than reactively. The Delphi method, for instance, taps into expert opinions to forecast potential outcomes and trends, while environmental scanning helps to identify external factors that could impact the research. In tandem, these methods allow researchers to craft informed hypotheses and set long-term goals that are aligned with anticipated future states. In today's world, where technological advancements and societal shifts occur at an accelerating pace, such foresight is not only beneficial but necessary for research to remain relevant and impactful. When combined, these techniques enable researchers to formulate well-informed hypotheses and establish long-term objectives that correspond with expected future conditions. Such foresight is not only advantageous but also required for research to remain relevant and meaningful in our modern world, where societal changes and technological developments happen at an accelerating rate.

Adopting adaptive research methodologies is equally important. Research is not conducted in a vacuum; it is influenced by changing stakeholder needs, unanticipated discoveries, and changing external situations. As a result, strict or overly limited approaches might quickly become out of date. Being adaptable is essential. As projects develop, researchers must use approaches that may grow or shrink to accommodate new deadlines, resources, and unanticipated challenges. This adaptability makes it possible for the research process to develop more naturally, allowing for modifications to be made without sacrificing the study's goals or overall integrity. In long-term research projects, where the original design may need to adapt significantly when new data or technologies become available, this kind of adaptability is very important.

Stakeholder engagement is another essential component of successful and sustainable research. It is important to involve stakeholders at every level of the research process, including funders, participants, policymakers, and community members. Stakeholders should indeed be included at every stage of the project's lifespan. This continuous involvement guarantees that the research stays in line with both short-term requirements and long-term goals. Continuous stakeholder involvement allows for the provision of insightful opinions, resources, and comments that can help to improve the research's direction and make it more responsive to demands in the real world. Furthermore, ongoing involvement builds confidence, strengthens collaborations, and raises the possibility that research findings will be successfully used or implemented.

Finally, *Iterative Development and Feedback Loops* provide a useful way to ensure that research is adaptable to new information and evolving conditions. Researchers can establish feedback loops that enable ongoing reflection and recalibration by incorporating iterative techniques within the research design. These loops serve as checkpoints where methods are adjusted in light of new information, data is examined, and theories are tested. In addition to ensuring accuracy, this iterative technique enables research to develop naturally, remaining pertinent to both short-term contexts and long-term trends. Iteration promotes creativity and lowers the danger of pursuing paths that may no longer be productive by encouraging experimentation and learning from mistakes.

When combined, these strategies provide a thorough foundation for conducting grounded and flexible research. Through trend forecasting, methodological adaptation, stakeholder engagement, and feedback loop construction, researchers can successfully negotiate the intricacies of the research process. These methods guarantee that research stays pertinent and aligned with present and future realities, producing more significant and long-lasting results.

The Function of Visions and Prospects in Design Research

In design-oriented research, the formulation of visions and prospects stands as a central pillar in guiding this form of scholarly work. In this context, prospects function as strategic tools that help researchers critically evaluate their direction and methodologies. They are a call to deliberate on the desirability of potential futures and to carefully consider the pathways that research might take toward those futures.

The function of prospects in HCI/interaction design research is also multifaceted. They offer a framework within which the feasibility and implications of research can be scrutinized. By articulating a clear vision of the desired outcome, prospects challenge researchers to assess their current trajectory critically. This involves asking a number of fundamental questions. For instance, are the research methodologies employed robust enough to lead to the envisioned outcomes? Do the methods allow for the adaptability required to navigate through evolving technological and societal landscapes? Is the research direction aligned with the ethical considerations and societal impacts that these future outcomes necessitate?

Such questions are fundamental because they set the contours for research projects' structure and execution. The choice of methods, the formulation of research questions, the analysis of data – all these elements of research design are influenced by the clarity and specificity of the prospects defined at the outset. Prospects that are well-articulated can serve to refine research questions, making them more precise and aligned with the desired outcomes. They can guide methodological choices, ensuring that the tools and techniques employed are appropriate for the depth and breadth of exploration required.

The articulation of clear research visions contributes to this process by supporting a culture of systematic knowledge production. Furthermore, such visions encourage researchers to approach inquiry with a mindset and attitude geared toward generating findings that hold value not only in the present but also related to a preferred future. As such, processes geared toward "knowing" are simultaneously about "nowing" in terms of changing the existing "now" into a more preferred future now. We see such efforts being made right now when researchers seek to address the UN's SDGs – Sustainability Development Goals, including efforts made for "the green transition" toward a more sustainable society. This is where the interplay between rigor and relevance becomes evident. Rigorous research is methodical, comprehensive, and grounded in robust scientific principles. However, without relevance, even the most rigorous research can become esoteric. Prospects ensure that the rigor applied in the research process yields outcomes pertinent to real-world scenarios, challenges, and opportunities. Here again, we cannot only rely on the importance (relevance) of the SDGs, but we also need basic research to rethink and reevaluate what we know today (knowledge and practices that brought us to the current situation), and reimagine and reorient our agenda to reach future goals. Here, a clear vision for the future is crucial to move from short-term actions to ways of working with future-distant goals (Feuls et al., 2024).

Moreover, prospects and visions work together, as a bridge, to connect disparate research efforts and align them with a shared research aim. In a field as dynamic and interdisciplinary as design, where research can range from highly theoretical explorations to applied technological development, prospects serve as unifying elements. They enable researchers from different domains and with various expertise to collaborate toward shared outcomes, fostering a transdisciplinary approach to problem-solving.

The development of visions and prospects is inherently a forward-thinking exercise. It projects the researcher into a future state where the consequences of today's research are lived and experienced. This projection is not merely aspirational. On the contrary, it is grounded in the present realities of the research environment and the societal context. It requires an awareness of current trends, challenges, and needs, as well as an anticipation of future shifts in knowledge, technology, and societal values.

In essence, the prospects and visions crafted and articulated by HCI/interaction design researchers act as commitments to a future they seek to make, enable, or influence. These are not passive actions but active constructions – a blending of anticipation and intention that shapes the direction of the research being conducted. As such, they are instrumental in ensuring that the research conducted is not only methodically sound but also holds significant potential for impact. By orienting research toward these carefully constructed prospects, researchers can align their work with both the present needs and future possibilities, making their contributions both timely and enduring.

MAKING TIME AND BREAKING NEW GROUNDS

In the field of design research, the value of research prospects to drive speculative design is increasingly recognized (see for instance Auger (2013), and Barendregt and Vaage (2021). These prospects, defined by their orientation toward future states, offer more than a mere outline of desired outcomes; they recalibrate the present, providing a lens through which the current state of research can be situated and aligned with future possibilities. This strategic alignment of research activities with a prospect of plausible or alternative future ensures that the investigative efforts made "now" are contributing to the landscape of an alternative "tomorrow." In a sense, this transition from now to the future, by undertaking actions now, to get to the future is something I would call "nowing" as an action-oriented complement to "knowing."

The development of such future-oriented research prospects necessitates a reassessment of current research trajectories. It's an invitation for scholars to pause and reflect critically on the path they are on, to evaluate the adequacy of existing methodologies, and to contemplate the need for novel approaches. This process of introspection for the individual researcher is crucial, for it is in this scrutiny that opportunities for innovation and growth are often found. The inquiry into whether current methods are sufficient to move us toward the envisioned future state can lead to a transformative realignment of research practices.

When we consider the formulation of research prospects as a recalibration of the present, we recognize that they serve as both a benchmark and a catalyst. They are benchmarks in that they establish a set of criteria against which the current state of research can be measured. How closely does ongoing research align with the envisioned outcomes? How does the work being conducted today bridge the gap between the present "now" and the anticipated future? As catalysts, research prospects spur change by challenging the contemporary moment and status quo, advocating for the development of new methodologies when existing ones fall short. Further on, they push the boundaries of traditional research paradigms, encouraging researchers to move into unexplored areas of research and embrace new, sometimes unconventional, modes of inquiry. If we look at the Nobel Prize winners, it is often this mindset that has given them the prize. It is not given to them because they have followed well-established methods and established ways of seeing things. On the contrary, it is because they have typically developed new methods, new approaches, and new ways of seeing things. By questioning what we think we know, we can also choose an alternative path for moving forward.

As such, prospects serve to extend the horizon of research, offering a forward-looking perspective that drives scholarly work. This perspective demands that researchers not only adapt to the currently unfolding future but also actively participate in its creation. In design research, this is sometimes described in terms of "futuring" (see e.g., Jenkins et al. (2024). and through methods that might include speculation Zhu et al. (2024b), fabulation Rosner (2018), or critical design approaches (see e.g., Bardzell et al., 2012).

Furthermore, the interplay between the present and the future is nuanced by the anticipation of challenges and the preemption of potential issues. Prospects empower researchers to forecast the needs and conditions of the future, incorporating an element of foresight into their work. This foresight is essential for crafting research that is resilient, flexible, and adaptable to the winds of change. In this context, prospects become more than static targets; they are dynamic elements of the research process, continuously shaped and reshaped by new findings, insights, and global developments.

The research strategy of "making time" through the process of formulating future prospects also implies a commitment to ongoing learning and an ability to adapt to changing circumstances – over time. As researchers engage with the current state of their field and the broader societal context, they must remain open to the idea that their understanding of the future – and the pathways leading there – will change, evolve, and might even be interrupted or completely changed. They must be willing to radically change and reevaluate, learn from emerging trends, and integrate new knowledge and new perspectives into their research plans. One example of this in the field of design research is the current and emerging strand of research on "more-than-human" design. From having developed a human-centered approach to design over the last four decades, we now see a complete re-orientation in our field where researchers seek to destabilize the human-centered approach, decentralize the human, and foreground other-than-human life forms (including plants, animals, and microbes). The idea is simple – we cannot only envision a future where humans live sustainable lives while still using the planet as a resource to produce food, energy for our homes, etc. Instead, we need to find alternative models where we build upon a relational understanding and where we as humans are just one part of a network of interrelated things in this world.

The process of working back and forth between the present state of research and its future prospects is a constant one, marked by the iterative assessment of progress and the continual refinement of research goals. In this extended discourse, researchers are not only investigators but also pioneers, breaking new ground as they explore new and alternative paths that might lead us toward a future we aspire to realize, whether it has to do with a "green transition," gender, social justice, or for instance ways of protecting personal integrity in the age of AI and the global harvesting of data. Through such processes, research prospects become integral to the scholarly journey, steering the course of inquiry, and underpinning the advancement of knowledge.

THE ROLE OF VISIONS AND PROSPECTS IN RESEARCH PROGRAMS

Formulating research visions employs scenarios, storytelling, and fabulation as tools to articulate, analyze, and extrapolate alternative futures (Nardi, 2015) and (Søndergaard et al., 2023, July). These narratives are essential in not just envisioning goals but in rendering the abstract tangible, providing a substrate for research objectives. In short, by telling stories about preferable futures, we can communicate our ideas about alternative futures, and through our scenarios and stories, others can see and participate in the making of these alternative future states.

In academic research, particularly within design-oriented disciplines, the establishment of a grand research vision can often seem overwhelming due to its scale and the complexity of achieving it. However, when this grand vision is deconstructed into smaller, actionable components, what was once daunting becomes attainable. These smaller components, or prospects, are essential for the formulation of long-term research programs that provide structure and guidance for a sequence of related research projects.

The significance of understanding and spending time on formulating the overarching prospect cannot be overstated. It is something that serves as the bedrock for the design of research programs. Such programs are akin to a master plan, encompassing a series of smaller research projects or work packages that collectively advance toward the overarching vision. Each project within the program is interconnected, contributing a piece of the puzzle, and is instrumental in gradually building a complete picture or a pathway to the future.

The development of long-term research programs requires strategic planning and forethought. It involves identifying the key milestones and outcomes that are necessary for the realization of the research vision. Prospects, as these manageable parts within the program, must be clearly defined, with research objectives that are both specific and measurable, or in other forms possible to evaluate. This clarity is what transforms a vision from an abstract idea into a concrete plan of action for a research program.

Research programs typically bring lots of people together to work on shared research problems and challenges over long periods of time. As such, these programs work as a hub for people with similar interests and research ambitions. These research programs can also serve as a roadmap, outlining the steps and stages of research from inception to completion. They allow researchers to prioritize their efforts, allocate resources effectively, coordinate efforts, and ensure that their work remains focused and cohesive. By breaking down the grand vision into smaller prospects, researchers can manage their tasks more effectively, making progress together in a systematic and organized manner – over longer periods of time. Moreover, these research programs provide a framework that can adapt to new discoveries and shifts in direction. As each project within the program progresses, researchers can reassess and refine their prospects in response to new information, ensuring that the research program remains relevant and on course to achieve the intended vision.

The long-term research programs are also not just about reaching an end goal; they are about fostering a culture of continuous inquiry and development. They encourage researchers to think beyond the confines of individual projects and consider how their work contributes to broader scholarly and societal advancements.

In designing these research programs, it's crucial to consider how each prospect fits into the larger vision. Researchers must ensure that there is a logical progression from one sub-project to the next, with each step building upon the last. This incremental progression is what enables significant advancements over time, as the culmination of numerous smaller achievements leads to the realization of the grand vision.

In essence, the role of prospects in research programs is to provide a structured approach to scholarly investigation and design exploration. They are instrumental in turning ambitious visions into practical plans, guiding researchers through the process of ideation, exploration, discovery, and innovation. Through careful planning and execution of these research programs, the grand visions of today can become the realities of tomorrow, and accordingly, it becomes an act of shaping the future of design-oriented research and its impact on the world.

ON MAKING RESEARCH CONTRIBUTIONS THROUGH VISIONS AND PROSPECTS

We have now been through a discussion on doing research "on time" vs. doing things that actually lead to ways of "making time" for timely research. This focus on conducting timely research is fundamentally connected to ways of making research contributions. Accordingly, I will, in this section relate this challenge of making research contributions to these future-oriented concepts of visions and prospects.

If we revisit the previous discussion in this book on fast vs. slow approaches to research, we can see how research in interaction design typically exists within a dual framework, where one axis is *reactive*, focusing on resolving immediate issues through the creation of products, while the other is *proactive*, exploring possibilities, and developing forward-looking prospects. Addressing problems with practical solutions is a significant part of interaction design research but is sometimes also criticized for exactly that reason under the label of "solutionism" (see e.g., Aanestad, 2023; Adamu, 2023). This approach involves understanding user needs and creating interactive systems and applications that can improve functionality and user experience. This is a pragmatic, solution-focused approach that has immediate applications in the real world.

On the other axis, the proactive axis, are prospects and visions that researchers formulate and cultivate for what could be. These are not just daydreams but carefully crafted projections of alternative future states, informed by research, data, and strategic foresight. Sometimes this process has been labeled "disciplined imagination" (ref), "foresight" (ref), or "future studies" (ref). In this context, the formulation of prospects draws from the well of possibilities, extrapolating current trends and knowledge into the formulation of comprehensive visions for the future.

Bridging Gaps through Prospects

Recalling earlier discussions from Chapter 2, prospects serve a dual purpose. They are not only endpoints or goals but also function as connectors, bridging the temporal divide between the immediacy of today's challenges and the potential for long-term societal impacts. They allow researchers to set long-range objectives that guide their inquiry, ensuring that each project contributes to a larger, coherent vision. In doing so, prospects merge the exploratory nature of basic research with the solution-driven focus of applied research.

Purpose and Direction in Research

Prospects provide purpose and direction to research activities. By defining clear future goals, they enable researchers to establish parameters for their inquiries, ensuring that the research remains aligned with long-term objectives while remaining flexible enough to adapt to new findings and changing conditions. This strategic approach to research planning ensures that every study, experiment, and analysis contributes to the broader vision.

Strategic Approach to Imaginative Foresight

Developing prospects requires a delicate balance of imagination and strategy. Researchers must think creatively about the future, identifying opportunities and risks that may not yet be evident. This imaginative foresight must be coupled with strategic planning to create pathways that are actionable and achievable. Researchers should consider the potential social, technological, and environmental changes that could influence their research, using this information to inform the development of their prospects.

Pathways from Possibilities to Prospects

Transforming possibilities into prospects is an exercise in strategic thinking and planning. Researchers must identify the steps necessary to move from the current state to the desired future state. This may involve the development of new technologies, the adoption of new methods, or the pursuit of new areas of inquiry. By understanding the journey from possibility to prospect, researchers can develop a roadmap for their research, detailing the steps needed to achieve their goals.

The prospects of design research are important because they provide a framework for understanding and evaluating the potential impact and relevance of a research project. Prospects can help to identify the potential contributions of a research project, as well as the potential challenges and limitations.

The formulation of research prospects is also, as I have described here, about envisioning a future. In return, these prospects enable us to situate ourselves in a possible/plausible future. This not only allows us to not only evaluate whether this future situation is preferred, but also serves as a guiding beacon for how we should navigate as we move forward in our research projects.

By considering the prospects of a research project, researchers can develop a more focused and well-defined research question, methodology, and plan. This can lead to more effective and impactful research, as well as a better understanding of the potential implications and applications of the research findings. The formulation of research visions and prospects is also beneficial because it breaks the temporally linear process of moving from a research problem and the identification of a research gap, to empirical studies, analysis, and conclusions, to a situation where we first need to think about where we are going with the research we do. This approach gives us the opportunity to critically examine whether we are approaching this future state, how far we are from that future state, and whether alternative methods and approaches are needed to get there.

Overall, the prospects of design research are crucial for understanding the potential value and significance of a research project. By carefully considering the prospects of a research project, researchers can conduct research that is both rigorous and relevant.

CONCLUSION

To summarize, this chapter emphasizes the dual roles that designers and researchers play – responding to the current cultural and technological contexts while simultaneously shaping the direction of future developments. Each decision, question, and innovation contributes to the ongoing evolution of design, and with it, the future of society. Designers are not passive observers but active participants in constructing this future through their work.

Further on, this chapter has emphasized the necessity of strategic alignment with current circumstances, which calls for alertness and adaptability in dealing with current issues. It also emphasizes how crucial it is to pursue long-term goals that transcend short-term trends. These goals ought to guarantee that design research is well-thought-out and progressive while also offering a framework for creativity. As we continue to engage in design research, there is an inherent responsibility to approach this role with purpose, recognizing the potential long-term impacts of our contributions. Our work extends beyond the documentation of design history; it serves as a foundation for future developments in various fields.

In this chapter, I have further elaborated on how it is crucial to approach research actively to stay effective, not only by following current trends but also by actively forming new ones – so-called "futuring" (see e.g., Bendor et al., 2021; Cornish, 2004; Fry, 2009). This strategy guarantees that our contributions will continue to be significant in the future, in the making of alternative futures, as well as being relevant in the present. One of the hallmarks of impactful design research is, accordingly, this continuous interplay between present and future situations. Researchers working in this area must, accordingly, address current issues while also planning for potential future developments, and strive for "future making." We can make sure that the design processes we develop today will contribute to the future in significant and long-lasting ways by firmly establishing our research in thoughtful investigations and explorations.

In the end, this chapter suggests that the act of HCI/interaction design research is inherently creative. It is through the deliberate formulation and articulation of visions and the careful crafting of prospects that design researchers can not only predict but actively shape the future. In this sense, our field is deeply future-oriented. Each question posed, each design exploration undertaken, and each innovation introduced is a building block in the construction of this future. HCI/interaction design researchers, therefore, are not just observers or commentators but creators, actively participating in the unfolding story of human knowledge and its applications. Thus, in asserting the indispensability of prospects and visions, this chapter calls upon HCI/interaction design researchers to approach their work with a future orientation, not only focusing on intellectual rigor and practical relevance. Beyond these two, a central concern is what future our efforts strive toward, and even enable.

In the next, and final chapter of this book, we will return to fast and slow as the overarching theme for this book, and in particular revisit the ideas of

- Urgency and scale

- Applied vs basic research

- Aim and approach

- Focus and frame

- Vision and prospects

And ways of seeing these ideas as elements of a thoughtful and integrated approach – across applied and basic research – to do timely and future-oriented research.

In addition, we will discuss why we, beyond these ideas, also need to constantly move between concrete and applied research projects, and more long-term and even eternal research questions. In short, we aim to meet the dual goal of addressing societal problems and needs, while advancing the foundational core of knowledge in the academic discipline.

Moving Forward – Toward the Future

INTRODUCTION

As we now arrive at the final chapter of this book, it is time to wrap things up and look to the future. This book took the problem of how to conduct timely design research as a central point of departure. From this position, I looked at fast vs. slow methods and approaches, and I considered how we sometimes need to move fast (maybe to follow how a certain digital technology is rapidly developing and spreading), and how we sometimes need to slow down, rethink, and apply a critical lens to examine and reexamine everything from a particular case to the assumptions we have made about the world.

This book has also related the fast vs. slow approaches to applied vs. basic research, and it has discussed ways of doing timely research where a combination of fast and slow approaches might be needed. Further on, it has addressed the scale and urgency of research problems, and it has discussed how this sets a frame for thinking about how much time we have for the research we set out to do. To paraphrase the American poet Amanda Gorman, – "There is always time, if we're only brave enough to take it, we're only brave enough to make it."[1]

Along these lines of "taking the time needed" or "making time" for the research you want to do, this final chapter also returns to the importance of having "eternal research questions," that is fundamental questions you

 DOI: 10.1201/9781003343745-9

return to over and over again throughout your research career. For me, it has been to deepen my understanding of what interaction is, how it might be supported or enabled through computational materials, and what new interactions unfold as a result of such digital designs.

In this book, I have also critically examined the fast and slow approaches, and finally, in the previous chapter, I asked if there was a way to also break the timeline we take so for granted when we do our research projects. For sure, we see it everywhere, with researchers trying to carefully plan their time, create timelines, set deadlines, formulate work packages, and visualize ways of pushing projects forward through GANTT charts for their projects. In every research grant application, the question of how the time in the project is managed is a central concern, and promised project deliverables are also typically marked with a deadline. So, given how occupied we are with these temporal concerns, what strategies can we develop to move forward?

This chapter proposes four such techniques for moving forward: 1) brainstorming, 2) historically rooted techniques – moving forward by looking back – historical studies, 3) prototyping futures – moving forward by imagining alternative futures, and 4) seeing with new eyes. These are a set of techniques that we can apply to generate ideas about future preferable states, but in moving forward we also need to think about what we mean when we say "moving forward." Are we moving forward from one research gap to the next? Or are we moving from the identification of a research gap to the formulation of a suitable research project to address that gap? Or are we moving forward in terms of scale, such as moving from small-scale research projects to larger research programs (longer time and more long-term goals)?

We can also think about moving forward in terms of progress and ways of moving forward in terms of moving from here to the future (in a linear or non-linear process), or ways to move toward the future and work backward (as we do with visions and as we do when we work back and forth in design research processes). We can also think about moving forward in terms of how we move from research programs to research prospects – to make plans for future research. Moving forward can also be about bringing the pieces together, collecting experiences from previous projects, and build motor themes, or it might be about building streams of research activities, papers, and collaborative networks.

In a sense, this thing about moving forward is in many ways related to your own motivation. How do you motivate your work? How do you connect the dots? And toward which envisioned future do you strive?

As you move forward, no matter what your personal motivation is, make sure you formulate a clear focus, something that you return to, your "course of action" or mission – and make sure to articulate your eternal research questions. Beyond research goals, aims, and visions, these eternal questions will guide you in moving forward.

Once you have identified your eternal questions, one way of moving forward is to make sure to always go back and forth between basic/fundamental/eternal questions and more applied questions. This gives you real-world oriented research projects, where you have shorter cycles of exploration, results, and impact, while it also fuels your more long-term research interests.

Given that you can move back and forth between these two states, you can create a certain rhythm to your HCI/interaction design research career where you move between two categories of inquiry: the applied questions, which look for quick, workable answers, and the fundamental questions, which address the timeless and universal. Establishing this back and forth rhythm takes careful navigation and understanding of the ways in which these two lines of research are connected and enhance one another in moving forward.

The eternal research questions are fundamental inquiries. These are the long-term questions that come up when we try to figure out, for instance "what interaction is really all about," "what the human experience is really about," and what our connections with digital technology are really all about. These eternal questions are typically "what and why" questions; they are frequently philosophical and can appear to have no answers. These questions should be viewed as the source from which applied questions gain their depth and significance rather than as esoteric diversions. These questions can give us a sense of purpose and anchor our work, ensuring that we are conducting meaningful investigations rather than just solving one issue after another. They serve as a reminder that human needs, cultural contexts, and temporal flows are at the core of every digital design.

In the context of sustainability and social justice – two topics frequently used in this book to exemplify more large-scale and long-term societal challenges – the research questions are not merely questions to be answered but challenges to be continually addressed. The pursuit of sustainability is a continuous process, a commitment to adapt and evolve practices and technologies in ways that can support a balance with our environment indefinitely. Similarly, social justice is an ongoing struggle, a series of actions and policies that must be constantly revisited and revised in pursuit of equity and fairness. However, only long-term thinking will not bring about any societal changes. So what about applied projects connected to these long-term ambitions?

Applied research projects act as stepping stones in this context. They are contextual, precise, and instantaneous. These are the "how" questions that call for our meticulousness, creative problem-solving, and flexibility in the face of a constantly evolving technological environment. Applied questions force us to think creatively within limitations, take customer wants and preferences as well as societal needs and challenges into account, and act quickly to satisfy them. They bring the abstract concepts of basic questions down to earth so they can be experimented with, refined, and made real in concrete ways.

The true depth of interaction design research is found in the interactions between these two categories of questions – between the eternal questions and the more short-term questions we also need to formulate to do something actionable. Progressing progressively entails having an ongoing conversation between the applied and the fundamental. It is an explorative process that acknowledges their interdependence rather than favoring one over the other. We refuel our knowledge and remind ourselves of the overarching topic, or the orientation of the research community we are a part of, when we go back to the basic questions.

To steadily move forward when shifting between eternal and more applied research questions is therefore about maintaining balance. It involves letting the applied questions guide and motivate the fundamental ones, and letting the applied work's discoveries help to clarify and occasionally even redefine our basic understanding. It's a cyclical dual process in which the degree to which we integrate the immediate with the eternal, rather than how far we go from the beginning, determines how far we've progressed.

Overall, if now concluding this exploration of ways of moving between eternal and more applied research questions, I would say that it is a philosophical question about time itself. What does it mean for research to be fast, slow, and timely? Is it about speed, about being first, or is it more about timing or the right audience for your "voice"? Timely research in design, then, is something that resonates with contemporary human needs and certain phases of societal development. For sure, eternal research questions have no value if never connected to the world around us, and vice versa, applied projects have little value if no knowledge is generalized into a form (typically a method or theory) that can be carried over to the next project.

But beyond these eternal questions and applied projects, I have in this book highlighted the importance of being future-oriented in the research agenda. To seek to formulate research aims that not only solve existing problems but also contribute to a better, or preferred, future. But how can we imagine such futures? What are the techniques we can use that help us in getting a glimpse of a preferable future? That is, what are the techniques we can apply for picturing this future state that we so desperately need as a thought figure for moving forward? In the beginning of this chapter, I proposes four such techniques for moving forward: 1) brainstorming, 2) historically rooted techniques – moving forward by looking back – historical studies, 3) prototyping futures – moving forward by imagining alternative futures, and 4) seeing with new eyes. In the following sections, we will accordingly take a closer look at these techniques for moving forward toward the future.

Techniques for Moving Forward

In the field of HCI/interaction design research, we have a set of techniques that we can apply to generate ideas about future preferable states. Over the years, our field of research has explored at least four different types of techniques: 1) brainstorming techniques, 2) historically rooted techniques, 3) future prototyping, and 4) ways of "seeing with new eyes."

1) Brainstorming techniques

Brainstorming might be the most common technique for generating and envisioning future states. Through brainstorming activities, new ideas are generated, shared, and discussed. This technique follows some simple rules, including how all ideas are allowed, and how no one is allowed to criticize ideas generated by someone else. The main focus of brainstorming sessions is to generate and share ideas – typically within a smaller group of people.

There are also additional, more specific techniques that have been developed for brainstorming, including, for instance the "six thinking hats" technique (de Bono, 1985). Here the idea is that it might be important to develop ideas along different types – to help in the exploration of a topic. Central to this approach is to avoid homogeneous thinking, and instead generate ideas along the themes of "information known or needed"/facts (the white hat), "brightness and optimism" (the yellow hat), "judgment"/the devil's advocate/difficulties and dangers (the black hat), "feelings, hunches, and intuition" (the red hat), "creativity" (the green hat), and the blue hat that serves as a control mechanism to manage this thinking process.

Through brainstorming, new ideas are generated that might then be closely examined, developed, and expanded upon. Through this technique, new ideas can be generated that open up for new paths for development and new opportunities.

2) Historically rooted techniques – Moving forward by looking back – historical studies

Another technique for moving forward is actually to look back in order to move forward. Through historical studies, design researchers can examine values, decisions, and structures that have influenced the development up until the current situation. The central idea here is that through an examination of these historical roots, one can understand much more about the current situation and also how a rethinking of dimensions and structures underpinning "the now" can help in imagining alternative futures.

This technique is about reading the current through the past, and it is about moving beyond the current by understanding the paths and the structures that have led up to the current situation.

3) Prototyping futures – Moving forward by imagining alternative futures

A third technique has to do with a method that is deeply rooted in design, that is to use prototyping at different scales. The most common approach is to use prototyping to try out new ideas in physical form – to get a sense of what it would be like if it turned into an object. However, prototyping is now increasingly used at additional scales – to sketch out what future workplaces, homes, and even societies could look like. While a vision might be instant and it might just be there in the beginning of a project, the refinement and articulation of the vision might take time – to put into words on what it is that the design should enable and to move the vision from idea to physical form. Prototyping is ideal for this transitional act.

Through prototyping, the design researcher can move forward through cycles of iterations, from an initial idea to a more informed understanding of how a particular design could enable or support a future preferred situation.

But prototyping is not limited to the construction of small-scale physical objects. Today, these prototypes can take the form of a formulated vision, a scenario, or a science fiction story that works as a vehicle for understanding what a future state could be. Prototyping can also happen through critical design studies where critical alternatives are articulated in the form of a prototype (see e.g., Bardzell et al., 2012). In short, prototyping can work on many different scales, and the prototype can be realized in many different forms – but it has one thing in common across any project that applies prototyping: it is to use the prototype to articulate and make visible the character of an alternative future.

4) Changing perspectives – "Seeing with new eyes"

The fourth technique has to do with moving forward by deliberately changing your perspectives – as to allow for seeing "with new eyes." If you are feeling stuck in your design research project or if you feel like you have been examining the current situation over and over again it might help to examine the grounds for your way of seeing – or put differently, through your theoretical and analytical lens. This is related to the "six thinking hats" technique (de Bono, 1985) with the idea of changing perspectives, but here it is not about applying one of these different hats, but more about critically examining the foundations through which you understand and "read" the world around you. Do you read it in terms of power relations? Or maybe in monetary terms? Do you look for injustices or inequalities? Or do you look for socio-technical interplays? Depending on your framing and way of seeing the world, you will see and notice different things. Accordingly, if you can make your own perspective clear to yourself, you can also start to think about alternative lenses for observing, seeing, or "reading" the world around you, and you will equip yourself with lenses for imagining alternative futures. Changing how we see things and shifting focus to other aspects changes not only our perception and understanding but also the things we envision. Moving forward through "new ways of seeing" or "seeing with new eyes" is about applying new theoretical perspectives, using alternative concepts to foreground what is important, to "read" the current situation, and imagine alternative futures by using alternative vocabularies to articulate these ideas. Changing perspectives is accordingly also about changing grounds. It is about altering one's perception of what is foregrounded and changing the perspective on how things are connected and interrelated. While brainstorming can be applied as a fairly quick approach to generate new ideas, it is likely that historical approaches, and to go deep into the underpinning perspectives, and even challenging these with alternative ways of seeing, takes longer time. These strategies complement each other, although they unfold along different temporalities.

From Eternal Questions and Techniques for Moving Forward to Ways of Integrating Basic Research and Applied Research in HCI Interaction Design Research

In the evolving field of human–computer interaction (HCI) and interaction design research, a central consideration is how the field progresses through the integration of basic and applied research. This integrative approach melds the theoretical and practical aspects of HCI to form a comprehensive view that informs both immediate design solutions and foundational theories.

Basic research in HCI often returns to the fundamental questions of "what is interaction?", "what is good/intended UX – user experiences" and "what is design?" These questions underpin the very essence of the field and ensure its continued relevance and adaptability to emerging technologies and societal changes. On the other hand, applied research seeks to resolve specific design problems or improve user experiences through practical applications. By interweaving these two research types, scholars can craft projects that both inform theory and translate directly into practice.

The implications for project timelines are significant. Short-term projects tend to prioritize applied research for its immediate impact, utilizing basic research as a supporting foundation. Conversely, long-term projects may focus on basic research to deepen understanding, with applied research translating these findings into real-world applications. This synergy propels the field toward a future where technology is more responsive to human needs and values.

The notion of "moving forward" in design-oriented research encompasses the transition from identifying research gaps to developing focused projects and, ultimately, to establishing expansive research programs. These programs represent concerted efforts to address grand challenges over time, aligning with broader goals such as sustainability and social justice – themes that are never fully resolved but require continuous commitment. Moreover, the linear progression from current research to future application is complemented by a visionary approach where the desired future state informs present research activities. This visioning process not only identifies what should be done but also backcasts to determine the steps necessary to achieve the envisioned future.

To foster sustained progress in HCI, researchers are encouraged to build thematic streams of research that contribute to broader discourses over time. These themes could range from the perpetual exploration of human-centered design principles to the burgeoning area of more-than-human-centered design. By connecting these thematic dots, researchers can set clear goals, establish networks, and create a ripple effect of knowledge that extends far beyond individual projects.

When thinking about the future of HCI research, one must consider the concept of eternal research questions – those that remain perpetually relevant and timely regardless of temporal trends. These enduring questions serve as a foundation upon which new research is built and provide a stable core for the field's progression. They reflect ongoing societal and environmental challenges that demand constant attention and effort, such as the quest for sustainable practices and the pursuit of social justice.

Integrating basic and applied research within interaction design research provides a thorough approach to understanding and developing the field. Basic research is the groundwork – it delves into the "whys" of human interaction with technology, laying the theoretical foundation upon which practical applications are built. Applied research then takes these theoretical frameworks and tests them in real-world scenarios, creating tangible products or solutions to everyday design problems.

In the context of a single project, integrating both types of research could involve an initial phase of intensive literature review and theory development. Scholars would pore over existing works, develop new frameworks or models, and hypothesize how these might manifest in practical settings. Subsequently, the project would shift into the application phase, where theories are operationalized through technology design, user testing, and iteration. This approach not only enriches the academic field with robust concepts but also benefits the industry with innovations grounded in solid research.

Alternatively, basic and applied research might progress on parallel tracks with strategic points of intersection. Fundamental research would be concerned with the conceptual expansion of interaction design, while applied research would concentrate on addressing current design challenges. The two streams would continuously inform each other – findings from basic research would guide the applied projects, ensuring that the solutions are not only innovative but also theoretically sound.

The implications of integrating basic and applied research are multifaceted when considering the duration of research projects. In short-term projects, the emphasis is likely on the applied side, generating solutions that can be quickly brought to market or implemented in practice. Here, basic research informs and refines the applied efforts, offering insight and depth that might otherwise be overlooked in the rush to deliver.

For long-term projects, the weight shifts toward basic research. These projects have the luxury of time, which allows for an in-depth exploration of interaction theories and the fundamental principles of design. Applied research in this context has a complementary role, gradually translating the insights gained from basic research into practical, long-term solutions that address evolving design challenges.

To summarize, I would say that integrating basic and applied research allows for an ongoing cycle of innovation, where theoretical advancements inspire new applications, and practical challenges stimulate further theoretical investigation. This symbiotic relationship is essential for the growth of interaction design, ensuring that the field remains dynamic, relevant, and responsive to both academic explorations and societal needs.

Moving Forward – Toward the Future

Beyond different techniques for moving forward and beyond any ambitions to integrate fast and slow approaches across applied and basic research, there needs to be an orientation for these efforts. In short, toward which future? And when we say "moving" then is it in terms of moving from one thing to the next? Is it about progress? Is it about getting closer to a solution or deeper understanding? Or is it more oriented toward further developments? It might be about moving from research gaps to research projects, moving from research projects to research programs (longer time, larger goals), or moving from here to the future (linear process), or to the future and back (by working with visions of a preferred future state). It might also be about moving from research programs to research prospects, building motor themes, and ways of building streams of research activities, papers, and networks. In this section, we take a closer look at these different ways of moving forward.

Conducting design-oriented research is a dynamic process that necessitates a progression from identifying what we don't know – research gaps – to creating structured initiatives – research projects – and eventually developing comprehensive research programs. This progression is vital to address complex challenges and to push the field forward.

Identifying research gaps is the first step toward meaningful contributions. It involves a critical analysis of the existing literature and practice to determine where the voids are – areas where questions remain unanswered, and problems unsolved. This gap analysis lays the groundwork for setting up research projects, which are targeted initiatives aimed at exploring these uncharted areas to generate new insights and solutions.

Moving from individual projects to research programs involves scaling up the focus. Research programs are assemblies of interconnected projects steered toward achieving broader, more ambitious goals. These programs are characterized by extended timelines, providing the space needed to tackle complex, long-standing issues that require sustained effort. The transition from projects to programs is strategic, often dictated by the overarching questions that drive the field and the long-term objectives of the research community.

The trajectory from the present state of research to future implications can typically be viewed as a linear process, where research builds upon itself over time. However, forward-thinking and future-oriented design research also involve a recursive process – envisioning future states and working backward to establish the steps necessary to realize those futures. This backcasting approach allows researchers to set milestones that guide their journey toward the envisioned outcomes.

To build momentum in research, themes – referred to here as motor themes – must be established. These are the central, driving questions or concepts that give direction to research activities. Motor themes provide coherence to the research efforts and are crucial for building streams of activities, which may include a series of papers, networking events, and collaborative projects. They serve as the thematic spine that supports the body of research work. Building these streams of research activities requires a deliberate effort to connect the dots across various studies, integrating findings and synthesizing knowledge. This interconnectedness ensures that the individual efforts contribute to a larger, cohesive understanding and that the research remains relevant to the field and its stakeholders.

Setting clear research goals is a necessary aspect of moving forward. Goals provide focus and direction, ensuring that research efforts are goal-oriented and purposeful. They help researchers to stay on course, measure progress, and evaluate the impact of their work.

Lastly, the act of envisioning and striving toward the future is fundamental to design-oriented research. It requires a visionary outlook – a willingness to imagine what could be and a commitment to the iterative process of research that leads to those future states. Prospects, or detailed projections of future outcomes, stem from this envisioning. They are the detailed maps that guide researchers toward the horizons of innovation and knowledge discovery.

In summary, moving forward in design-oriented research is an active, iterative process. It involves recognizing gaps, initiating projects, expanding to programs, envisioning futures, and systematically working toward them through interconnected research activities and networks. By connecting the present with future aspirations, researchers ensure that their work is both impactful today and will continue to resonate and provide value in the years to come.

Concluding Reflection – On Fast and Slow – Over Time

In the end, we might need to accept that in the social sciences, and in design-oriented research, there is no absolute truth. And accordingly, research is never finished – we will never arrive at "the truth" and have a research problem solved once and for all. Instead, our research efforts are a continuum where we continuously explore, seek new paths, revisit old paths, and through these efforts strive toward a better future for all.

So maybe research isn't after all about the (final) results, the outputs we produce, the timelines we follow, or the deadlines we meet. Maybe it is not about the goal or the final destination, but rather about the path and the process. This turns research away from a production machine and more into a practice, where we continuously practice (as in *learning*), and where we continuously practice (as in *doing* research).

Then, ultimately, research is about *learning* and *doing* – reflecting and making – and how you do it, and along which paces you do it, is your decision as an independent researcher.

But also, as I have pinpointed in this book, we do not only conduct research as independent researchers. We do it together, collectively, in the research community. We do it through our "academic citizenship," through the projects we run together, through the questions we ask, through the prototypes we build and evaluate, and through the models, methods, and theories we develop. We also conduct research for our surrounding society. In short, we do it together, and we do it for each other.

However, there is sometimes this misconception that we do research for someone "out there" in "society." This is a very vague formulation and a very vaguely addressed audience for our research. Of course, we should be glad if someone picks up our research, but mostly we are doing it for the development of the research area and for the research community.

While consultants might need to "fix" problems for someone else, researchers have a different task at hand. This task is less about fixing something for someone, but rather about contributing to ways of thinking together about certain things. It might be how to understand something out there in the world (a phenomenon), how to approach it (a method), or how to articulate its character, mechanisms, and behavior (conceptual or theoretical work). Of course, all of this might be useful for someone "out there" in "society," but it is equally important to understand that independent research might be important enough for an open, democratic society – to have organizations such as universities where researchers can think freely, seek new knowledge, and question what we think we know. This is important for modern society.

So, given that we are not just serving society by "solving societal challenges," we can give ourselves the time and space to also freely decide how we want to approach the research challenges we see. Do you want to go slow, or do you prefer to go fast? That is what I have kept as a central concern across the chapters in this book.

For sure, it might be hard to just go fast – as it will probably lead to a sense of lacking foundational grounding. And it might be equally hard to go slow – as it will probably lead to a sense of missing how the world changes around us. Accordingly, and given that research needs to be a free activity, it is ultimately a personal choice how to combine these two approaches in your research practice.

Coming back to a central message in this book – a combination of fast and slow enables a dual focus on constantly changing societal needs and new emerging technologies while making space for more long-term theoretical work that can help to describe and explain the world around us. Still, while there might be urgent and large-scale problems to tackle, the way you combine fast and slow approaches in your scholarly work is ultimately about your personal motivation and how you want to be heard – in the research community and in our surrounding society. Given that, these questions about how to conduct timely research, how to make timely contributions, and how to choose a suitable research methodology are ultimately about your "voice" and thus ultimately about *you*.

Timely research is conducted by those who have a clear voice, by those who have decided how they want to be heard, and accordingly – and beyond any fast and slow approaches – it is these voices that form our field of research – now and for times to come.

NOTE

1. The original phrasing from her speech at the Presidential inauguration for President Joe Biden was: – "There is always light, if we're only brave enough to see it, if we're only brave enough to be it."

References

Aanestad, M. (2023, December). Digital degrowth–beyond solutionism. In *IFIP joint working conference on the future of digital work: The challenge of inequality* (pp. 55–60). Springer Nature Switzerland.

Acquaviva, G. J. (2000). Experiencing interconnectedness. In *Values, violence, and our future* (pp. 153–175). Brill.

Adams, A., & Cox, A. L. (2008). *Questionnaires, in-depth interviews and focus groups* (pp. 17–34). Cambridge University Press.

Adamu, M. S. (2023). No more "solutionism" or "saviourism" in futuring African HCI: A manyfesto. *ACM Transactions on Computer-Human Interaction, 30*(2), 1–42.

Ahlqvist, T., & Uotila, T. (2020). Contextualising weak signals: Toward a relational theory of futures knowledge. *Futures, 119*, 102543.

Åhlström, P., & Karlsson, C. (2010). Longitudinal field studies. In *Researching operations management* (pp. 210–249). Routledge.

Akerman, N. (2018). *The necessity of friction.* Routledge.

Alavi, H., Churchill, E., Wiberg, M., Lalanne, D., Dalsgaard, P., & Fatah, A. (2019). Special Issue on Human-Building Interaction ACM Transactions on Computer-Human Interaction (TOCHI), ACM. Introduction to human-building interaction (HBI): Interfacing HCI with architecture and urban design. *ACM Transactions on Computer-Human Interaction (TOCHI), 26*(2), 6.

Alexander, P. A., Graham, S., & Harris, K. R. (1998). A perspective on strategy research: Progress and prospects. *Educational Psychology Review, 10*, 129–154.

Alleva, L. (2006). Taking time to savour the rewards of slow science. *Nature, 443*(7109), 271–271.

Arnowitz, J., Arent, M., & Berger, N. (2010). *Effective prototyping for software makers.* Elsevier.

Arvola, M., Forsblad, M., Wiberg, M., & Danielsson, H. (2023). Autonomous vehicles for children with mild intellectual disability. In European Conference in Cognitive Ergonomics (ECCE '23), September 19–22, 2023, Swansea, United Kingdom. ACM, New York, NY.

Ashby, S., Hanna, J., Matos, S., Nash, C., & Faria, A. (2019, November). Fourth-wave HCI meets the 21st century manifesto. In *Proceedings of the halfway to the future symposium 2019* (pp. 1–11).

Auger, J. (2013). Speculative design: Crafting the speculation. *Digital Creativity, 24*(1), 11–35.

Avison, D. E., Lau, F., Myers, M. D., & Nielsen, P. A. (1999). Action research. *Communications of the ACM, 42*(1), 94–97.

Ball, L. J., Christensen, B. T., & Halskov, K. (2021). Sticky notes as a kind of design material: How sticky notes support design cognition and design collaboration. *Design Studies, 76*, 101034.

Bardzell, J., Bardzell, S., Dalsgaard, P., Gross, S., & Halskov, K. (2016, June). Documenting the research through design process. In *Proceedings of the 2016 ACM conference on designing interactive systems* (pp. 96–107).

Bardzell, S., Bardzell, J., Forlizzi, J., Zimmerman, J., & Antanitis, J. (2012, June). Critical design and critical theory: the challenge of designing for provocation. In *Proceedings of the designing interactive systems conference* (pp. 288–297).

Barendregt, L., & Vaage, N. S. (2021). Speculative design as thought experiment. *She Ji: The Journal of Design, Economics, and Innovation, 7*(3), 374–402.

Bascur, C., Rusu, C., & Quiñones, D. (2019). User as customer: touchpoints and journey map. In *Human systems engineering and design: Proceedings of the 1st International Conference on Human Systems Engineering and Design (IHSED2018): Future trends and applications, October 25–27, 2018, CHU-Université de Reims Champagne-Ardenne, France 1* (pp. 117–122). Springer International Publishing.

Bastian, M., Jones, O., Moore, N., & Roe, E. (Eds.). (2016). *Participatory research in more-than-human worlds*. Routledge.

Bates, O., Thomas, V., Remy, C., Nathan, L. P., Mann, S., & Friday, A. (2018, April). The future of HCI and sustainability: Championing environmental and social justice. In *Extended abstracts of the 2018 CHI conference on human factors in computing systems* (pp. 1–4).

Bauman, Z. (2013). *Liquid modernity*. John Wiley & Sons.

Beck, J., & Bergqvist, E. S. (2018, November 5). Designing has no given problems, no given processes, and no given solutions. https://doi.org/10.31219/osf.io/dz9xr

Bell, G., & Dourish, P. (2007). Yesterday's tomorrows: Notes on ubiquitous computing's dominant vision. *Personal and Ubiquitous Computing, 11*, 133–143.

Bendor, R., Eriksson, E., & Pargman, D. (2021). Looking backward to the future: On past-facing approaches to futuring. *Futures, 125*, 102666.

Benford, S., Giannachi, G., Koleva, B., & Rodden, T. (2009, April). From interaction to trajectories: Designing coherent journeys through user experiences. In *Proceedings of the SIGCHI conference on human factors in computing systems* (pp. 709–718).

Benyon, D. (2014). *Designing interactive systems: A comprehensive guide to HCI, UX and interaction design*. Pearson.

Berg, M., & Seeber, B. K. (2016). *The slow professor: Challenging the culture of speed in the academy*. University of Toronto Press.

Bergström, J., & Hornbæk, K. (2025). DIRA: A model of the user interface. *International Journal of Human-Computer Studies, 193*, 103381.

Berkovich, I. (2020). Something borrowed, something blue: Reflections on theory borrowing in educational administration research. *Journal of Educational Administration, 58*(6), 749–760.

Bertrand, J. T., Brown, J. E., & Ward, V. M. (1992). Techniques for analyzing focus group data. *Evaluation Review, 16*(2), 198–209.

Besant, H. (2016). The journey of brainstorming. *Journal of Transformational Innovation, 2*(1), 1–7.

Bibri, S. E. (2018). Backcasting in futures studies: A synthesized scholarly and planning approach to strategic smart sustainable city development. *European Journal of Futures Research, 6*(1), 1–27.

Blackwell, A. F. (2015, April). HCI as an inter-discipline. In *Proceedings of the 33rd annual ACM conference extended abstracts on human factors in computing systems* (pp. 503–516).

Blandford, A., Furniss, D., & Makri, S. (2016). *Qualitative HCI research: Going behind the scenes.* Morgan & Claypool.

Bley, J., Eriksson, A., Johansson, L., & Wiberg, M. (2023). Design friction in autonomous drive—exploring transitions between autonomous and manual drive in non-urgent situations. *Personal and Ubiquitous Computing, 27*(6), 2291–2305.

Blythe, M. (2014, April). Research through design fiction: Narrative in real and imaginary abstracts. In *Proceedings of the SIGCHI conference on human factors in computing systems* (pp. 703–712).

Blythe, M., Andersen, K., Clarke, R., & Wright, P. (2016, May). Anti-solutionist strategies: Seriously silly design fiction. In *Proceedings of the 2016 CHI conference on human factors in computing systems* (pp. 4968–4978).

Boring, E. G. (1955). Dual role of the Zeitgeist in scientific creativity. *The Scientific Monthly, 80*(2), 101–106.

Bottero, W. (2009). Relationality and social interaction 1. *The British Journal of Sociology, 60*(2), 399–420.

Bozalek, V. (2017). Slow scholarship in writing retreats: A diffractive methodology for response-able pedagogies1. *South African Journal of Higher Education, 31*(2), 40–57.

Brooke, J. (1996). *SUS: A quick and dirty usability scale.* Usability Evaluation in Industry.

Buchanan, R. (1992). Wicked problems in design thinking. *Design Issues, 8*(2), 5–21.

Burkitt, I. (1997). The situated social scientist: Reflexivity and perspective in the sociology of knowledge. *Social Epistemology, 11*(2), 193–202.

Bødker, S. (2006, October). When second wave HCI meets third wave challenges. In *Proceedings of the 4th Nordic conference on Human-computer interaction: Changing roles* (pp. 1–8).

Bødker, S. (2015). Third-wave HCI, 10 years later—participation and sharing. *Interactions, 22*(5), 24–31.

Camburn, B., Viswanathan, V., Linsey, J., Anderson, D., Jensen, D., Crawford, R., & Wood, K. (2017). Design prototyping methods: State of the art in strategies, techniques, and guidelines. *Design Science, 3*, e13.

Camocini, B., & Vergani, F. (2021). *From human-centered to more-than-human-design: Exploring the transition* (p. 161). FrancoAngeli.

Carp, J. (2011). *The study of slow*. MIT.

Carrol, J. M. (1999, January). Five reasons for scenario-based design. In *Proceedings of the 32nd annual hawaii international conference on systems sciences. 1999. hicss-32. abstracts and cd-rom of full papers* (p. 11). IEEE.

Carroll, J. M. (2003). *Making use: Scenario-based design of human-computer interactions*. MIT Press.

Chang, Y. N., Lim, Y. K., & Stolterman, E. (2008, October). Personas: From theory to practices. In *Proceedings of the 5th Nordic conference on Human-computer interaction: Building bridges* (pp. 439–442).

Checkland, P., & Scholes, J. (1999). *Soft systems methodology in action*. John Wiley & Sons.

Chordia, I., Baltaxe-Admony, L. B., Boone, A., Sheehan, A., Dombrowski, L., Le Dantec, C. A., & Smith, A. D. (2024, May). in HCI: A systematic literature review. In *Proceedings of the CHI conference on human factors in computing systems* (pp. 1–33).

Clark, D. (2021). *The long game: How to be a long-term thinker in a short-term world*. Harvard Business Press.

Cohen, L., Manion, L., & Morrison, K. (2017). Action research. In *Research methods in education* (pp. 440–456). Routledge.

Collins, A., Joseph, D., & Bielaczyc, K. (2016). Design research: Theoretical and methodological issues. In *Design-based research* (pp. 15–42). Psychology Press.

Cooper, C. M. (2024). Design timescapes: Futuring through visual thinking. *Visual Communication, 23*(1), 172–188.

Cornish, E. (2004). *Futuring: The exploration of the future*. World Future Society.

Coulton, P., & Lindley, J. G. (2019). More-than-human centred design: Considering other things. *The Design Journal, 22*(4), 463–481.

Croskerry, P., Petrie, D. A., Reilly, J. B., & Tait, G. (2014). Deciding about fast and slow decisions. *Academic Medicine, 89*(2), 197–200.

Cunningham, J. J., & MacEachern, S. (2016). Ethnoarchaeology as slow science. *World Archaeology, 48*(5), 628–641.

Cunningham, J., Benabdallah, G., Rosner, D., & Taylor, A. (2023). On the grounds of solutionism: Ontologies of blackness and HCI. *ACM Transactions on Computer-Human Interaction, 30*(2), 1–17.

Dalton, N. S., Schnädelbach, H., Wiberg, M., & Varoudis, T. (2016). *Architecture and interaction*. Springer.

de Bono, E. (1985). Six thinking hats. Penguin Books.

de La Bellacasa, M. P. (2017). *Matters of care: Speculative ethics in more than human worlds* (Vol. 41). University of Minnesota Press.

De Sá, M., & Carriço, L. (2006, April). Low-fi prototyping for mobile devices. In *CHI'06 extended abstracts on Human factors in computing systems* (pp. 694–699).

Di Poppa, F. (2010). Spinoza and process ontology. *The Southern Journal of Philosophy, 48*(3), 272–294.

DiSalvo, C., Lukens, J., Lodato, T., Jenkins, T., & Kim, T. (2014, April). Making public things: How HCI design can express matters of concern. In *Proceedings of the SIGCHI conference on human factors in computing systems* (pp. 2397–2406).

DiSalvo, C., Sengers, P., & Brynjarsdóttir, H. (2010, April). Mapping the landscape of sustainable HCI. In *Proceedings of the SIGCHI conference on human factors in computing systems* (pp. 1975–1984).

Dombrowski, L., Harmon, E., & Fox, S. (2016, June). Social justice-oriented interaction design: Outlining key design strategies and commitments. In *Proceedings of the 2016 ACM conference on designing interactive systems* (pp. 656–671).

Dourish, P. (2006). Implications for design. In *Proceedings of the SIGCHI conference on human factors in computing systems* (pp. 541–550).

Dourish, P. (2010, August). HCI and environmental sustainability: the politics of design and the design of politics. In *Proceedings of the 8th ACM conference on designing interactive systems* (pp. 1–10).

Dow, S., MacIntyre, B., Lee, J., Oezbek, C., Bolter, J. D., & Gandy, M. (2005). Wizard of Oz support throughout an iterative design process. *IEEE Pervasive Computing*, 4(4), 18–26.

Earley, R. (2017). Designing fast & slow. Exploring fashion textile product lifecycle speeds with industry designers. *The Design Journal*, 20(supl), S2645–S2656.

Erete, S., Rankin, Y. A., & Thomas, J. O. (2021). I can't breathe: Reflections from Black women in CSCW and HCI. *Proceedings of the ACM on Human-Computer Interaction*, 4(CSCW3), 1–23.

Erete, S., Rankin, Y., & Thomas, J. (2023). A method to the madness: Applying an intersectional analysis of structural oppression and power in HCI and design. *ACM Transactions on Computer-Human Interaction*, 30(2), 1–45.

Ertner, M., Kragelund, A. M., & Malmborg, L. (2010, November). Five enunciations of empowerment in participatory design. In *Proceedings of the 11th biennial participatory design conference* (pp. 191–194).

Escobar, A. (2011). Sustainability: Design for the pluriverse. *Development*, 54, 137–140.

Escobar, A. (2018). *Designs for the pluriverse: Radical interdependence, autonomy, and the making of worlds*. Duke University Press.

Feuls, M., Hernes, T., & Schultz, M. (2024). Putting distant futures into action: How actors sustain a course of action toward distant-future goals through path enactment. *Academy of Management Journal*, (ja), amj-2022.

Forlano, L. E., & Halpern, M. K. (2023). Speculative histories, just futures: From counterfactual artifacts to counterfactual actions. *ACM Transactions on Computer-Human Interaction*, 30(2), 1–37.

Fox, S., Asad, M., Lo, K., Dimond, J. P., Dombrowski, L. S., & Bardzell, S. (2016, May). Exploring social justice, design, and HCI. In *Proceedings of the 2016 CHI conference extended abstracts on human factors in computing systems* (pp. 3293–3300).

Frauenberger, C. (2019). Entanglement HCI the next wave? *ACM Transactions on Computer-Human Interaction (TOCHI)*, 27(1), 1–27.

Frith, U. (2020). Fast lane to slow science. *Trends in Cognitive Sciences, 24*(1), 1–2.

Frohlich, D. M. (2015). *Fast design, slow innovation.* Springer International Publishing.

Fruehwald, J. (2017). Generations, lifespans, and the zeitgeist. *Language Variation and Change, 29*(1), 1–27.

Fry, T. (2009). *Design futuring* (pp. 71–77). University of New South Wales Press.

Gasparin, M., Green, W., & Schinckus, C. (2020). Slow design-driven innovation: A response to our future in the Anthropocene epoch. *Creativity and Innovation Management, 29*(4), 551–565.

Ghajargar, M., Bardzell, J., Renner, A. S., Krogh, P. G., Höök, K., Cuartielles, D., & Wiberg, M. (2021, February). From "explainable ai" to "graspable ai". In *Proceedings of the fifteenth international conference on tangible, embedded, and embodied interaction* (pp. 1–4).

Gawande, A. (2013). Slow ideas. *The New Yorker, 29*, 36–45.

Gazzaley, A., & Rosen, L. D. (2016). *The distracted mind: Ancient brains in a high-tech world.* MIT Press.

Giaccardi, E. (2019). Histories and futures of research through design: From prototypes to connected things. *International Journal of Design, 13*(3), 139–155.

Giaccardi, E., Redström, J., & Nicenboim, I. (2024). The making (s) of more-than-human design: Introduction to the special issue on more-than-human design and HCI. *Human–Computer Interaction, 40*, 1–16.

Glass, C. R., & Fitzgerald, H. E. (2010). Engaged scholarship: Historical roots, contemporary challenges. In *Handbook of engaged scholarship: Contemporary landscapes, future directions* (pp. 9–24). Michigan State University Press.

Godin, D., & Zahedi, M. (2014). Aspects of research through design: A literature review. In *Proceedings of the design research society conference 2014,* Umeå, Sweden, 16–19 June 2014.

Goh, C. H., Kulathuramaiyer, N., & Zaman, T. (2017). Riding waves of change: A review of personas research landscape based on the three waves of HCI. In *Information and communication technologies for development: 14th IFIP WG 9.4 international conference on social implications of computers in developing countries, ICT4D 2017, Yogyakarta, Indonesia, May 22–24, 2017, Proceedings 14* (pp. 605–616). Springer International Publishing.

Goldsworthy, K., Earley, R., & Politowicz, K. (2018). Circular speeds: A review of fast & slow sustainable design approaches for fashion & textile applications. *Journal of Textile Design Research and Practice, 6*(1), 42–65.

Gough, B., & Madill, A. (2012). Subjectivity in psychological science: From problem to prospect. *Psychological Methods, 17*(3), 374.

Grandia, L. (2015). Slow ethnography: A hut with a view. *Critique of Anthropology, 35*(3), 301–317.

Grover, V., & Lyytinen, K. (2015). New state of play in information systems research. *MIS Quarterly, 39*(2), 271–296.

Handwerker, W. P. (2001). *Quick ethnography.* Rowman Altamira.

Haraway, D. J. (2016). Staying with the trouble: Making kin in the Chthulucene. In *Staying with the trouble.* Duke University Press.

Harding, S. (2008). *Sciences from below: Feminisms, postcolonialities, and modernities.* Duke University Press.

Harrington, C., Rosner, D., Taylor, A., & Wiberg, M. (2021). Engaging race in HCI. *Interactions, 28*(5), 5–5.

Hartman, Y., & Darab, S. (2012). A call for slow scholarship: A case study on the intensification of academic life and its implications for pedagogy. *Review of Education, Pedagogy, and Cultural Studies, 34*(1–2), 49–60.

Hartson, H. R. (1998). Human–computer interaction: Interdisciplinary roots and trends. *Journal of Systems and Software, 43*(2), 103–118.

Hayward, S. (2016). *Fast and slow: Design and the experience of time.* Middlesex University.

Holland, D., Powell, D. E., Eng, E., & Drew, G. (2010). Models of engaged scholarship: An interdisciplinary discussion. *Collaborative Anthropologies, 3*(1), 1–36.

Hollway, W. (2010). Preserving vital signs: The use of psychoanalytically informed interviewing and observation in psycho-social longitudinal research. In *Intensity and insight: Qualitative longitudinal methods as a route to the psycho-social,* 19. Working Paper.

Holmlid, S. (2009, November). Participative, co-operative, emancipatory: From participatory design to service design. In *First Nordic conference on service design and service innovation* (Vol. 53).

Holmquist, L. E. (2017). Intelligence on tap: Artificial intelligence as a new design material. *Interactions, 24*(4), 28–33.

Homewood, S., Karlsson, A., & Vallgårda, A. (2020, July). Removal as a method: A fourth wave HCI approach to understanding the experience of self-tracking. In *Proceedings of the 2020 ACM designing interactive systems conference* (pp. 1779–1791).

Houseman, M. (2006). Relationality. In *Theorizing rituals, volume 1: Issues, topics, approaches, concepts* (pp. 413–428). Brill.

Hughes, J., King, V., Rodden, T., & Andersen, H. (1995). The role of ethnography in interactive systems design. *Interactions, 2*(2), 56–65.

Hyysalo, S., Kohtala, C., Helminen, P., Mäkinen, S., Miettinen, V., & Muurinen, L. (2014). Collaborative futuring with and by makers. *CoDesign, 10*(3–4), 209–228.

Höök, K., & Löwgren, J. (2012). Strong concepts: Intermediate-level knowledge in interaction design research. *ACM Transactions on Computer-Human Interaction (TOCHI), 19*(3), 1–18.

Issa, T., & Isaias, P. (2022). Usability and Human–Computer Interaction (HCI). In *Sustainable design: HCI, usability and environmental concerns* (pp. 23–40). Springer London.

Janlert, L. E., & Stolterman, E. (2017). *Things that keep us busy: The elements of interaction.* MIT Press.

Jenkins, T., Tsaknaki, V., Howell, N., Boer, L., Wong, R. Y., Campo Woytuk, N., & Søndergaard, M. L. J. (2024, July). Mapping futures and futuring in HCI/design. In *Companion publication of the 2024 ACM designing interactive systems conference* (pp. 458–461).

Jones, J. C. (1992). *Design methods.* John Wiley & Sons.

Jones, T. S., & Richey, R. C. (2000). Rapid prototyping methodology in action: A developmental study. *Educational Technology Research and Development,* 48(2), 63–80.

Jung, H., Wiltse, H., Wiberg, M., & Stolterman, E. (2017). Metaphors, materialities, and affordances: Hybrid morphologies in the design of interactive artifacts. *Design Studies, 53,* 24–46.

Jungnickel, K. (Ed.). (2020). *Transmissions: Critical tactics for making and communicating research.* MIT Press.

Justa, B., Martins, N., Pereira, L., & Brandão, D. (2024). Waste management: Designing a user-centered and sustainable mobile application. In *Perspectives on design III: Research, education and practice* (pp. 77–93). Springer Nature Switzerland.

Kahneman, D. (2011). *Thinking, fast and slow.* Farrar, Straus and Giroux.

Kamrani, A. K., & Nasr, E. A. (2010). *Engineering design and rapid prototyping.* Springer Science & Business Media.

Kannengiesser, U., & Gero, J. S. (2019). Design thinking, fast and slow: A framework for Kahneman's dual-system theory in design. *Design Science, 5,* e10.

Kaptelinin, V., Nardi, B., Bødker, S., Carroll, J., Hollan, J., Hutchins, E., & Winograd, T. (2003, April). Post-cognitivist HCI: Second-wave theories. In *CHI'03 extended abstracts on Human factors in computing systems* (pp. 692–693).

Keulemans, G. (2021). *New materials: Fast tech and slow design.* The University of South Australia.

Kim, W. C., & Mauborgne, R. (2011). *Blue ocean strategy.* Harvard Business Review.

Kim, W. C., & Mauborgne, R. A. (2014). *Blue ocean strategy, expanded edition: How to create uncontested market space and make the competition irrelevant.* Harvard Business Review Press.

Kjærup, M., Skov, M. B., Nielsen, P. A., Kjeldskov, J., Gerken, J., & Reiterer, H. (2021). *Longitudinal studies in HCI research: A review of CHI publications from 1982–2019* (pp. 11–39). Springer International Publishing.

Knoblauch, H. (2005). Focused ethnography. In *Forum qualitative sozialforschung/forum: qualitative social research* (Vol. 6, No. 3).

Knowles, B., Bates, O., & Håkansson, M. (2018, April). This changes sustainable HCI. In *Proceedings of the 2018 CHI Conference on human factors in computing systems* (pp. 1–12).

Kochanowska, M., & Gagliardi, W. R. (2022). The double diamond model: In pursuit of simplicity and flexibility. In D. Raposo, J. Neves, & J. Silva (Eds.), *Perspectives on design II: Research, education and practice* (pp. 19–32). Springer International Publishing.

Kozubaev, S., Elsden, C., Howell, N., Søndergaard, M. L. J., Merrill, N., Schulte, B., & Wong, R. Y. (2020, April). Expanding modes of reflection in design futuring. In *Proceedings of the 2020 CHI conference on human factors in computing systems* (pp. 1–15).

Kozubaev, S., Elsden, C., Howell, N., Søndergaard, M. L. J., Merrill, N., Schulte, B., & Wong, R. Y. (2020, April). Expanding modes of reflection in design futuring. In *Proceedings of the 2020 CHI conference on human factors in computing systems* (pp. 1–15).

Krause, M. (2019). What is Zeitgeist? Examining period-specific cultural patterns. *Poetics, 76*, 101352.

Kuhn, T. (1970). *The nature of scientific revolutions.* University of Chicago.

Kuniavsky, M. (2009). User experience and HCI. In *Human-computer interaction* (pp. 19–38). CRC Press.

Kuznetsov, S., Hudson, S. E., & Paulos, E. (2014, February). A low-tech sensing system for particulate pollution. In *Proceedings of the 8th international conference on tangible, embedded and embodied interaction* (pp. 259–266).

Lewis, J. R. (2012). Usability testing. In G. Salvendy (Ed.), *Handbook of human factors and ergonomics* (pp. 1267–1312). John Wiley and Sons.

Linehan, C., Kirman, B. J., Reeves, S., Blythe, M. A., Tanenbaum, T. J., Desjardins, A., & Wakkary, R. (2014). Alternate endings: using fiction to explore design futures. In *CHI'14 Extended Abstracts on Human Factors in Computing Systems,* (pp. 45–48).

Linklater, A. (2010). Human interconnectedness. In K. Booth (Ed.), *Realism and world politics* (pp. 313–329). Routledge.

Lopes, A. G. (2021, May). HCI four waves within different interaction design examples. In *IFIP working conference on human work interaction design* (pp. 83–98). Springer International Publishing.

Lönngren, J., & Van Poeck, K. (2021). Wicked problems: A mapping review of the literature. *International Journal of Sustainable Development & World Ecology, 28*(6), 481–502.

Löwgren, J., & Stolterman, E. (2004). *Thoughtful interaction design.* MIT Press.

Lowgren, J., & Stolterman, E. (2007). *Thoughtful interaction design: A design perspective on information technology.* MIT Press.

Lucero, A., Desjardins, A., & Neustaedter, C. (2021). Longitudinal first-person HCI research methods. In E. Karapanos, J. Gerken, J. Kjeldskov, & M. B. Skov (Eds.), *Advances in longitudinal HCI research* (pp. 79–99). Springer.

Lutz, J. F. (2012). Slow science. *Nature Chemistry, 4*(8), 588–589.

MacDonald, C. M., & Atwood, M. E. (2013). Changing perspectives on evaluation in HCI: Past, present, and future. In *CHI'13 extended abstracts on human factors in computing systems* (pp. 1969–1978).

Mackay, W. E., & Fayard, A. L. (1997, August). HCI, natural science and design: a framework for triangulation across disciplines. In *Proceedings of the 2nd conference on Designing interactive systems: Processes, practices, methods, and techniques* (pp. 223–234).

Marila, M. M. (2019). Slow science for fast archaeology. *Current Swedish Archaeology, 27*(1), 93–114.

Mathiassen, L., & Nielsen, P. A. (2008). Engaged scholarship in IS research. *Scandinavian Journal of Information Systems, 20*(2), 1.

McCracken, G. (1988). *The long interview.* Sage Publications.

McKelvey, B. (2006). Van De Ven and Johnson's "engaged scholarship": Nice try, but…. *Academy of Management Review, 31*(4), 822–829.

Mencarini, E., Bremer, C., Leonardi, C., Liu, J., Nisi, V., Nunes, N. J., & Soden, R. (2023, April). HCI for climate change: Imagining sustainable futures. In *Extended abstracts of the 2023 CHI conference on human factors in computing systems* (pp. 1–6).

Meyerhoff, E., & Noterman, E. (2019). Revolutionary scholarship by any speed necessary: Slow or fast but for the end of this world. *ACME: An International Journal for Critical Geographies, 18*(1), 217–245.

Millen, D. R. (2000, August). Rapid ethnography: Time deepening strategies for HCI field research. In *Proceedings of the 3rd conference on Designing interactive systems: Processes, practices, methods, and techniques* (pp. 280–286).

Mitrović, I., Auger, J., Hanna, J., & Helgason, I. (2021). *Beyond speculative design: Past–present–future*. University of Split.

Mountz, A., Bonds, A., Mansfield, B., Loyd, J., Hyndman, J., Walton-Roberts, M., & Curran, W. (2015). For slow scholarship: A feminist politics of resistance through collective action in the neoliberal university. *ACME: An International Journal for Critical Geographies, 14*(4), 1235–1259.

Myers, B. (1994). Challenges of HCI design and implementation. *Interactions, 1*(1), 73–83.

Nardi, B. (2015). Designing for the future: But which one?. *Interactions, 23*(1), 26–33.

Nardi, B. A. (1997). The use of ethnographic methods in design and evaluation. In *Handbook of human-computer interaction* (pp. 361–366). North-Holland.

Neeley Jr, W. L., Lim, K., Zhu, A., & Yang, M. C. (2013, August). Building fast to think faster: Exploiting rapid prototyping to accelerate ideation during early stage design. In *International design engineering technical conferences and computers and information in engineering conference* (Vol. 55928, p. V005T06A022). American Society of Mechanical Engineers.

Nelson, H. G., & Stolterman, E. (2014). *The design way: Intentional change in an unpredictable world*. MIT Press.

Newell, A., & Card, S. K. (1985). The prospects for psychological science in human-computer interaction. *Human-Computer Interaction, 1*(3), 209–242.

Newport, C. (2016). *Deep work: Rules for focused success in a distracted world*. Hachette UK.

Nielsen, J. (1992). Finding usability problems through heuristic evaluation. In *Proceedings of the SIGCHI conference on human factors in computing systems* (pp. 373–380).

Nielsen, J. (1993). Iterative user-interface design. *Computer, 26*(11), 32–41.

Nielsen, J. (1994). *Usability engineering*. Morgan Kaufmann.

Nijs, G., Laki, G., Houlstan, R., Slizewicz, G., & Laureyssens, T. (2020, October). Fostering more-than-human imaginaries: Introducing DIY speculative fabulation in civic HCI. In *Proceedings of the 11th Nordic Conference On Human-computer Interaction: Shaping experiences, shaping society* (pp. 1–12).

Nordin, H., Almeida, T., & Wiberg, M. (2023). Designing to restory the past: Storytelling for empowerment through a digital archive. *International Journal of Design, 17*(1), 91–104.

Ogbonnaya-Ogburu, I. F., Smith, A. D., To, A., & Toyama, K. (2020, April). Critical race theory for HCI. In *Proceedings of the 2020 CHI conference on human factors in computing systems* (pp. 1–16).

Oomen, J., Hoffman, J., & Hajer, M. A. (2022). Techniques of futuring: On how imagined futures become socially performative. *European Journal of Social Theory, 25*(2), 252–270.

Ormiston, G. L., & Schrift, A. D. (Eds.). (1990). *The hermeneutic tradition: From ast to ricoeur.* Suny Press.

Oulasvirta, A., & Hornbæk, K. (2016, May). HCI research as problem-solving. In *Proceedings of the 2016 CHI conference on human factors in computing systems* (pp. 4956–4967).

Owens, B. (2013a). Long-term research: Slow science. *Nature, 495*(7441), 300–303.

Owens, B. (2013b). Slow science: The world's longest-running experiments remind us that science is a marathon, not a sprint. *Nature, 495*(7441), 300–304.

Packer, B. W., & Keates, S. (2024). HCI design methods. In *Designing for usability, inclusion and sustainability in human-computer interaction* (pp. 26–77). CRC Press.

Pelto, P. J. (2016). *Applied ethnography: Guidelines for field research.* Routledge.

Petersmann, M. C. (2021). Response-abilities of care in more-than human worlds. In *Posthuman legalities* (pp. 102–124). Edward Elgar Publishing.

Picho, K., Maggio, L. A., & Artino, A. R. (2016). Science: The slow march of accumulating evidence. *Perspectives on Medical Education, 5*, 350–353.

Pink, S., & Morgan, J. (2013). Short-term ethnography: Intense routes to knowing. *Symbolic Interaction, 36*(3), 351–361.

Poikolainen Rosén, A., Normark, M., & Wiberg, M. (2022). Toward more-than-human-centred design: Learning from gardening. *International Journal of Design, 16*(3), 21–36.

Putman, V. L., & Paulus, P. B. (2009). Brainstorming, brainstorming rules and decision making. *The Journal of Creative Behavior, 43*(1), 29–40.

Querejazu, A. (2016). Encountering the pluriverse: Looking for alternatives in other worlds. *Revista Brasileira de Política Internacional, 59*(2), e007.

Qutoshi, S. B. (2018). Phenomenology: A philosophy and method of inquiry. *Journal of Education and Educational Development, 5*(1), 215–222.

Raita, E. (2012, October). User interviews revisited: Identifying user positions and system interpretations. In *Proceedings of the 7th Nordic conference on human-computer interaction: Making sense through design* (pp. 675–682).

Randall, D., Harper, R., & Rouncefield, M. (2007). Ethnography and how to do it. In D. Randall, R. Harper, & M. Rouncefield (Eds.), *Fieldwork for design: Theory and practice* (pp. 169–197). Springer.

Rapp, A., Odom, W., Pschetz, L., & Petrelli, D. (2022). Introduction to the special issue on time and HCI. *Human–Computer Interaction, 37*(1), 1–14.

Redstrom, J. (2017). *Making design theory.* MIT Press.

Rittel, H. W., & Webber, M. M. (1973). Dilemmas in a general theory of planning. *Policy Sciences, 4*(2), 155–169.

Robles, E., & Wiberg, M. (2010, January). Texturing the" material turn" in interaction design. In *Proceedings of the fourth international conference on Tangible, embedded, and embodied interaction* (pp. 137–144).

Robles, E., & Wiberg, M. (2011). From materials to materiality: Thinking of computation from within an Icehotel. *Interactions, 18*(1), 32–37.

Rogers, Y. (2012). *HCI theory: Classical, modern, and contemporary* (Vol. 14). Morgan & Claypool Publishers.

Rosenbaum, S., Cockton, G., Coyne, K., Muller, M., & Rauch, T. (2002, April). Focus groups in HCI: Wealth of information or waste of resources?. In *CHI'02 extended abstracts on human factors in computing systems* (pp. 702–703).

Rosner, D. K. (2018). *Critical fabulations: Reworking the methods and margins of design.* MIT Press.

Rosner, D. K., Wiberg, M., & Taylor, A. S. (2021). The labor of tech. *Tech Labor, 28,* 4.

Rosner, D., Taylor, A., Wiberg, M., & Windle, A. (2021). The urgency for access. *Interactions, 28*(3), 5–5.

Rossitto, C., Comber, R., Tholander, J., & Jacobsson, M. (2022, April). Toward digital environmental stewardship: The work of caring for the environment in waste management. In *Proceedings of the 2022 CHI conference on human factors in computing systems* (pp. 1–16).

Rozanski, E. P., & Haake, A. R. (2003, October). The many facets of HCI. In *Proceedings of the 4th conference on information technology curriculum* (pp. 180–185).

Rudd, J., Stern, K., & Isensee, S. (1996). Low vs. high-fidelity prototyping debate. *Interactions, 3*(1), 76–85.

Ryd, N. (2004). The design brief as carrier of client information during the construction process. *Design Studies, 25*(3), 231–249.

Saad, E., Elekyaby, M. S., Ali, E. O., & Hassan, S. F. A. E. (2020). Double diamond strategy saves time of the design process. *International Design Journal, 10*(3), 211–222.

Salo, P., & Heikkinen, H. L. (2018). Slow science: Research and teaching for sustainable praxis. *Confero, 6*(1), 87–111.

Sauer, H. (2018). *Moral thinking, fast and slow.* Routledge.

Savin-Baden, M., & Niekerk, L. V. (2007). Narrative inquiry: Theory and practice. *Journal of Geography in Higher Education, 31*(3), 459–472.

Schlesinger, A., Edwards, W. K., & Grinter, R. E. (2017, May). Intersectional HCI: Engaging identity through gender, race, and class. In *Proceedings of the 2017 CHI conference on human factors in computing systems* (pp. 5412–5427).

Schmidt, F. L. (1992). What do data really mean? Research findings, meta-analysis, and cumulative knowledge in psychology. *American Psychologist, 47*(10), 1173.

Schön, D. A. (1971). *Beyond the stable state: Public and private learning in a changing society.* Penguin.

Schön, D. A. (2017). *The reflective practitioner: How professionals think in action*. Routledge.

Scotchmer, S. (1991). Standing on the shoulders of giants: Cumulative research and the patent law. *Journal of Economic Perspectives, 5*(1), 29–41.

Sefelin, R., Tscheligi, M., & Giller, V. (2003, April). Paper prototyping-what is it good for? A comparison of paper-and computer-based low-fidelity prototyping. In *CHI'03 extended abstracts on Human factors in computing systems* (pp. 778–779).

Sharma, V., Kumar, N., & Nardi, B. (2023). Post-growth human–computer interaction. *ACM Transactions on Computer-Human Interaction, 31*(1), 1–37.

Shneiderman, B. (2000). Universal usability. *Communications of the ACM, 43*(5), 84–91.

Simon, H. A. (2019). *The sciences of the artificial, reissue of the third edition with a new introduction by John Laird*. MIT Press.

Skaburskis, A. (2008). The origin of "wicked problems". *Planning Theory & Practice, 9*(2), 277–280.

Smith, D. G. (1991). Hermeneutic inquiry: The hermeneutic imagination and the pedagogic text. In E. C. Short (Ed.), *Forms of curriculum inquiry* (p. 3). SUNY Press.

Soden, R., Pathak, P., & Doggett, O. (2021, June). What we speculate about when we speculate about sustainable HCI. In *Proceedings of the 4th ACM SIGCAS conference on computing and sustainable societies* (pp. 188–198).

Søndergaard, M. L. J., Campo Woytuk, N., Howell, N., Tsaknaki, V., Helms, K., Jenkins, T., & Sanches, P. (2023, July). Fabulation as an approach for design futuring. In *Proceedings of the 2023 ACM designing interactive systems conference* (pp. 1693–1709).

Sørensen, C. (1994). This is not an article. Just some food for thought on how to write one. In *17th Information systems Research seminar* (pp. 46–59). University of Oulu.

Stengers, I. (2016). "Another science is possible!": A plea for slow science. In *Demo (s)* (pp. 53–70). Brill.

Stephanidis, C., Salvendy, G., Antona, M., Chen, J. Y., Dong, J., Duffy, V. G., & Zhou, J. (2019). Seven HCI grand challenges. *International Journal of Human–Computer Interaction, 35*(14), 1229–1269.

Stolterman, E., & Wiberg, M. (2010). Concept-driven interaction design research. *Human–Computer Interaction, 25*(2), 95–118.

Sturdee, M., & Lindley, J. (2019, November). Sketching & drawing as future inquiry in HCI. In *Proceedings of the halfway to the future symposium 2019* (pp. 1–10).

Sum, C. M., Alharbi, R., Spektor, F., Bennett, C. L., Harrington, C. N., Spiel, K., & Williams, R. M. (2022, April). Dreaming disability justice in HCI. In *CHI conference on human factors in computing systems extended abstracts* (pp. 1–5).

Sutton, R. I., & Hargadon, A. (1996). Brainstorming groups in context: Effectiveness in a product design firm. *Administrative Science Wuarterly, 41*, 685–718.

Svetel, I., Kosić, T., & Pejanović, M. (2018). Digital Vs. traditional design process. In *Proceedings of 5th international academic conference on places and technologies, "places and technologies 2018–keeping up with technologies to adapt cities for future challenges"* (pp. 453–460). Faculty of Architecture.

Swanson, D., Goel, L., Francisco, K., & Stock, J. (2017). Applying theories from other disciplines to logistics and supply chain management: A systematic literature review. *Transportation Journal, 56*(3), 299–356.

Tan, J. (2023, July). "User journeys": A tool to align cross-functional teams. In *International conference on human-computer interaction* (pp. 163–167). Springer Nature Switzerland.

Thinyane, M., Bhat, K., Goldkind, L., & Cannanure, V. K. (2018, August). Critical participatory design: Reflections on engagement and empowerment in a case of a community based organization. In *Proceedings of the 15th participatory design conference: Full papers-volume 1* (pp. 1–10).

Thomas, V., Remy, C., Hazas, M., & Bates, O. (2017, May). HCI and environmental public policy: Opportunities for engagement. In *Proceedings of the 2017 CHI conference on human factors in computing systems* (pp. 6986–6992).

Tripp, S. D., & Bichelmeyer, B. (1990). Rapid prototyping: An alternative instructional design strategy. *Educational Technology Research and Development, 38*(1), 31–44.

Twidale, M. B., Weber, N., Chamberlain, A., Cunningham, S. J., & Dix, A. (2014, February). Quick and dirty: Lightweight methods for heavyweight research. In *Proceedings of the companion publication of the 17th ACM conference on computer supported cooperative work & social computing* (pp. 343–346).

Tynan, L. (2021). What is relationality? Indigenous knowledges, practices and responsibilities with kin. *Cultural Geographies, 28*(4), 597–610.

Ulmer, J. B. (2017). Writing slow ontology. *Qualitative Inquiry, 23*(3), 201–211.

Upcraft, S., & Fletcher, R. (2003). The rapid prototyping technologies. *Assembly Automation, 23*(4), 318–330.

Valentin, D., Chollet, S., Lelièvre, M., & Abdi, H. (2012). Quick and dirty but still pretty good: A review of new descriptive methods in food science. *International Journal of Food Science & Technology, 47*(8), 1563–1578.

Verlie, B. (2022). Climate justice in more-than-human worlds. *Environmental Politics, 31*(2), 297–319.

Vindrola-Padros, C., & Vindrola-Padros, B. (2018). Quick and dirty? A systematic review of the use of rapid ethnographies in healthcare organisation and delivery. *BMJ Quality & Safety, 27*(4), 321–330.

Wakkary, R. (2021). *Things we could design: For more than human-centered worlds*. MIT Press.

Walker, M., Takayama, L., & Landay, J. A. (2002, September). High-fidelity or low-fidelity, paper or computer? Choosing attributes when testing web prototypes. In *Proceedings of the human factors and ergonomics society annual meeting* (Vol. 46, No. 5, pp. 661–665). Sage Publications.

Wania, C. E., Atwood, M. E., & McCain, K. W. (2006, June). How do design and evaluation interrelate in HCI research?. In *Proceedings of the 6th conference on Designing Interactive systems* (pp. 90–98).

Weick, K. E. (1989). Theory construction as disciplined imagination. *Academy of Management Review, 14*(4), 516–531.

White, R. T., & Arzi, H. J. (2005). Longitudinal studies: Designs, validity, practicality, and value. *Research in Science Education, 35,* 137–149.

Whittaker, S., Terveen, L., & Nardi, B. A. (2000). Let's stop pushing the envelope and start addressing it: A reference task agenda for HCI. *Human–Computer Interaction, 15*(2–3), 75–106.

Wiberg, M. (2010). *Interactive textures for architecture and landscaping: Digital elements and technologies.* IGI Global.

Wiberg, M. (2001, September). RoamWare: An integrated architecture for seamless interaction in between mobile meetings. In *Proceedings of the 2001 ACM international conference on supporting group work* (pp. 288–297).

Wiberg, M. (2010). Interactive architecture as digital texturation: Transformed public spaces & new material integration. In *Industrial informatics design, use and innovation: Perspectives and services* (pp. 44–57). IGI Global.

Wiberg, M. (2012). Interaction per se: Understanding "the ambience of interaction" as manifested and situated in everyday & ubiquitous IT-use. In *Innovative applications of ambient intelligence: Advances in smart systems* (pp. 71–97). IGI Global Scientific Publishing.

Wiberg, M. (2013). Making the case for "architectural informatics": A new research horizon for ambient computing?. In *Pervasive and ubiquitous technology innovations for ambient intelligence environments* (pp. 128–135). IGI Global.

Wiberg, M. (2014a). Interaction design research and the future. *Interactions, 21*(2), 22–23.

Wiberg, M. (2014b). Methodology for materiality: Interaction design research through a material lens. *Personal and Ubiquitous Computing, 18,* 625–636.

Wiberg, M. (2018). *The materiality of interaction: Notes on the materials of interaction design.* MIT Press.

Wiberg, M. (2020). On physical and social distancing: Reflections on moving just about everything online amid Covid-19. *Interactions, 27*(4), 38–41.

Wiberg, M. (2023). Critically analyzing autonomous materialities. In *Handbook of critical studies of artificial intelligence* (pp. 737–748). Edward Elgar Publishing.

Wiberg, M., & Robles, E. (2010). Computational compositions: Aesthetics, materials, and interaction design. *International Journal of Design, 4*(2), 65–76.

Wiberg, M., & Stolterman, E. (2014, October). What makes a prototype novel? A knowledge contribution concern for interaction design research. In *Proceedings of the 8th Nordic conference on human-computer interaction: Fun, fast, foundational* (pp. 531–540).

Wiberg, M., & Stolterman, E. (2019). Philosophy, HCI, and 'thought styles'. In *CHI´19 (Conference on Human Factors in Computing Systems)" Weaving the threads of CHI",* Glasgow, Scotland, UK, May 4–9, 2019. ACM Press.

Wiberg, M., & Stolterman, E. (2021). Time and temporality in HCI research. *Interacting with Computers, 33*(3), 250–270.

Wiberg, M., & Stolterman, E. (2023). Automation of interaction—interaction design at the crossroads of user experience (UX) and artificial intelligence (AI). *Personal and Ubiquitous Computing, 27*(6), 2281–2290.

Wiberg, M., & Whittaker, S. (2005). Managing availability: Supporting lightweight negotiations to handle interruptions. *ACM Transactions on Computer-Human Interaction (TOCHI), 12*(4), 356–387.

Wiberg, M., Taylor, A., & Rosner, D. (2022). Climate care. *Interactions, 29*(1), 5–5.

Wiberg, M., & Teigland, R. (2024, October). Computing for the 22nd century: More-than-human to see the environmental footprints of profound technologies. In *Proceedings of the halfway to the future symposium* (pp. 1–4).

Wilkinson, S. (1998). Focus group methodology: A review. *International Journal of Social Research Methodology, 1*(3), 181–203.

Williamson, T. (2019). Armchair philosophy. *Epistemology & Philosophy of Science, 56*(2), 19–25.

Wilson, C. (2013). *Brainstorming and beyond: A user-centered design method.* Newnes.

Wilson, J., & Rosenberg, D. (1988). Rapid prototyping for user interface design. In *Handbook of human-computer interaction* (pp. 859–875). North-Holland.

Wong, R. Y., & Khovanskaya, V. (2018). *Speculative design in HCI: From corporate imaginations to critical orientations* (pp. 175–202). Springer International Publishing.

Xu, W. (2019). Toward human-centered AI: A perspective from human-computer interaction. *Interactions, 26*(4), 42–46.

Yan, X., & Gu, P. E. N. G. (1996). A review of rapid prototyping technologies and systems. *Computer-Aided Design, 28*(4), 307–318.

Yoo, D., Bekker, T., Dalsgaard, P., Eriksson, E., Fougt, S. S., Frauenberger, C., & Wiberg, M. (2023, April). More-than-human perspectives and values in human-Computer Interaction. In *Extended abstracts of the 2023 CHI conference on human factors in computing systems* (pp. 1–3).

Zhu, L., Chao, C., & Fu, Z. (2024a). How HCI integrates speculative thinking to envision futures. *Journal of Futures Studies, 28*(4), 89–103.

Zhu, L., Wang, J., & Li, J. (2024b). Exploring the roles of artifacts in speculative futures: Perspectives in HCI. *Systems, 12*(6), 194.

Zilber, T. B. (2015). Studying organizational fields through ethnography. In *Handbook of qualitative organizational research* (pp. 86–96). Routledge.

Zimmerman, J., & Forlizzi, J. (2014). Research through design in HCI. In *Ways of knowing in HCI* (pp. 167–189). Springer New York.

Zimmerman, J., Forlizzi, J., & Evenson, S. (2007, April). Research through design as a method for interaction design research in HCI. In *Proceedings of the SIGCHI conference on Human factors in computing systems* (pp. 493–502).

Öhlund, L., & Wiberg, M. (2025). Social Justice in HCI: Current Streams, Considerations, and Ways Forward, Interacting with Computers, iwaf009, https://doi.org/10.1093/iwc/iwaf009

Index

For Product Safety Concerns and Information please contact our EU
representative GPSR@taylorandfrancis.com
Taylor & Francis Verlag GmbH, Kaufingerstraße 24, 80331 München, Germany